Springer Biographies

More information about this series at http://www.springer.com/series/13617

Wolfgang W. Osterhage

Galileo Galilei

At the Threshold of the Scientific Age

 Springer

Wolfgang W. Osterhage
Wachtberg, Nordrhein-Westfalen
Germany

ISSN 2365-0613   ISSN 2365-0621 (electronic)
Springer Biographies
ISBN 978-3-030-06297-2   ISBN 978-3-319-91779-5 (eBook)
https://doi.org/10.1007/978-3-319-91779-5

Printed on acid-free paper

This Springer imprint is published by the registered company Springer International Publishing AG
part of Springer Nature
The registered company address is: Gewerbestrasse 11, 6330 Cham, Switzerland

Preface

Why another book about Galileo? To most readers, his achievements, his life, and his conflict with the church authorities are well known. This book, however, tries to put the complex personality of Galileo into a different perspective. Although it still follows the stations of his life in their historical sequence, it relates his discoveries and theories to today's scientific developments in major fields of physics on the one hand and to ancient teachings on the other hand. Galileo's most important publications are discussed in detail.

The book starts with a broad description of major historical events during Galileo's lifetime, followed by a recounting of his early years in Pisa and his gravitational experiments there. During his time in Padua, he made his first important astronomical discoveries, which led to a first conflict with the church authorities later during his stay in Florence. At the centre of this conflict—as is well known—was his interest in the Copernican world model.

At this stage, a general outline is given of the development of cosmological world models from the ancient Greeks over Ptolemy and Copernicus up to our Standard Big Bang Model. The climax of Galileo's conflict was triggered by the publication of his "Dialogo." The narrative of the book concludes with a summary of his work in his final years in semi-exile in the countryside near Florence, followed by an assessment of the impact of his work for science in general and today's outlook in particular.

At the end, an exhaustive timetable containing the most important historical and scientific events including those of Galileo's life is presented. There is also a table of references used in the book. Most of the letters referred to in the text and some of the passages from his publications have been taken from A. Mudry, *Galileo Galilei: Schriften, Briefe, Dokumente*, Berlin, 1985, and retranslated by myself. Other facts about Galileo's life and circumstances can be found under Galileo.rice.edu.

My thanks go to Springer International Publishing for making this book possible in the first place. Special thanks are due to Dr. Angela Lahee for her patience, Ute Heuser, and Stephen Pfeiffer for his critical language review.

Wachtberg, Germany   Wolfgang W. Osterhage
May 2018

Contents

Chapter 1
Introduction

The life of Galileo, one of the founding fathers of modern science, is full of genius, but also of discrepancies, inconsistencies and retractions all of which had already been noticed and documented by his contemporaries. Despite his designation as the "Columbus of the Skies" they were well aware that he was a devout Catholic throughout his life. And he was not alone. In Italy at his time but also in Germany, more than one scholar was forced to wear a mask to hide his true beliefs in public.

Of course, there exists a whole panoply of biographies of this illustrious personality, in which the reader can find everything he wants to look for: his scientific achievements, his academic stations, his family life and last but not least his monumental conflict with the church establishment. So why bother to write another one?

The present book tries to deviate from the classical path of following the life of a prominent personality of historical fame and painstakingly reporting the details of his existence. In fact this is not its objective, although it is catalysed by a biography as such, but it is more a biography of science or rather of specific fields of natural science, of physics in particular. But it becomes evident in its course that even specialist scientific areas like astronomy very quickly stretch out their tentacles into other related subjects of interest such as mechanics or thermodynamics.

The figure of Galileo offers itself as a perfect example for the transition that took place during his lifetime all over Europe—be it political, religious or scientific. Galileo serves as a sort of burning glass, by which the rays of many ideas emanating during his lifetime were bundled and focussed. This book is thus not so much a biography of a single person but represents a tale of an excerpt of human history starting off as early as 490 BC and leaving off in 2012, as can be seen from the time line in Chap. 11. And somewhere in between along this line appears the man whose life serves as the leitmotif of this tale.

Of course, the milestones in this narrative are the personal dates and achievements of Galileo, his academic career, his scientific reflexions and his quarrels with the ancient philosophers. Important events include his major publications and discoveries, which enriched the view of the world, not only for his contemporaries, and his inventions. Already during his lifetime, legend building began, and fame surrounded his name even before his death.

© Springer International Publishing AG, part of Springer Nature 2018
W. W. Osterhage, *Galileo Galilei*, Springer Biographies,
https://doi.org/10.1007/978-3-319-91779-5_1

But the foundation of Galileo's fame was not only built on the sober results of his scientific work. One of the main reasons for remembering him more than others is to be found in his conflict with the church. Although he was far from the only person even of his time to have suffered this course of events, his fate had been exploited from the very beginning and is still up to today. This famous conflict is not the key subject of this book, although a whole chapter is devoted to it, but rather the cause of the conflict, i.e. the publication of a book that deals with two competing world models.

The Greek language provides two different translations for our word "time": Chronos and Kairos. Chronos refers to our usual understanding of time as consisting of a sequence of consecutive events to be measured by clocks. Kairos means time fulfilled, accomplished, consummated. The time was ripe at the threshold from the Middle Ages to modern times: Kairos. In fact the whole spiritual and philosophical (in its widest sense as comprising natural philosophy) climate generated this "threshold", and Galileo stands for the symbolic crossing of that threshold.

If we look at the mirror of the time in question, then that time was imprinted with the concurrence of the asynchronous. The church was the real point of reference: it controlled the life of the individual from birth via baptism and marriage up to death. At the same time, it wielded worldly powers. People of the outgoing 16th century were pious as never before or thereafter. At the same time, it was common understanding that man was at the centre of the cosmos.

The time was ripe. The central question that arose and would thereafter never leave the scholarly discussion until today dealt with our position in the world and not only in the cosmos as such. Are we Copernican, i.e. do we believe that man does in the end not occupy a preferred position in the world in any sense of meaning? Or is there still something non-Copernican, which may assign a special place for us in the universe? This was the question in a nutshell taken up in Galileo's Dialogo, and this is a question still in contention today.

The Anthropic Principle

As one key element in the dialogue between natural science and the theology of creation, the so-called fine adjustment is cited again and again, i.e. the ideal conditions allowing for the development of human life as such. The fine adjustment of our living space is indeed also the subject of an internal discussion among the sciences that puts the purely statistical arguments about infinitely possible living environments in the cosmos into perspective. In this connection, some time ago an epistemological basic debate was triggered in natural science and in its discourse with theology [1].

The fundamental question thereby is: why are we in a position to discern the world the way it is, and thus are able to describe it within the framework of a comprehensive theory?

If, for example, the neutron-proton mass difference were different by just a tiny amount, there could be not nuclear physics in the present sense, nor could there be elements or stars. Or: if the energy level in the C^{12}-nucleus did not correspond to 7.65 MeV, there would be no life based on carbon chemistry.

It looks as if many natural constants are just situated within the very narrow bounds that enable human life. This fact is called "the Anthropic Principle". Opinions diverge about its meaning. These are some:

- Since we are here and observe everything, those parameters just have to be as they are. Otherwise, we would not be there and could not wonder about them.
- Thus Life is something extremely improbable and something very special.
- The universe had been created in just this way to enable life.
- Rubbish.

To delve more deeply into this complex question, let us start with the three slights suffered by mankind:

- Transition from the geocentric to the heliocentric world model; Earth and man are no longer at the centre of the cosmos (Copernicus, Galileo): this means that we have to accept a progressive decentralisation and de-anthropomorphisation. The sun is just one star among many, our Milky Way just one galaxy among many, our cosmos—who knows—maybe only one (fractional) cosmos among many.
- The special status of man among living creatures is doubtful; mechanisms have replaced purpose (Darwin, Monod).
- Man no longer "rules the roost": the unconscious governs to a high degree (Freud). Recently this has been complemented by evolutionary cognitive science, artificial intelligence, and robot technology.

Against this backdrop, there are the following considerations:

On Earth, there is a life form with a consciousness, an observing intelligence. What must the corresponding universe look like? This question cannot be answered without the following logical steps:

- Consciousness presupposes that there is life.
- Life needs chemical elements as a precondition to come into existence, especially those that are heavier than hydrogen and helium.
- Heavy elements can only originate by thermonuclear reactions with light elements—nuclear fusion.
- Nuclear fusion takes place only in stars and needs at least several billion years to produce significant numbers of heavy elements.
- The time span of several billion years is only at the disposal in a universe which itself is several billion years old and has been extended by several billion light years.
- On the other hand, in later cosmic epochs stars similar to our sun very rarely would come into existence, but mainly white dwarfs with low energies unable to supply a planetary, slowly evolving life form with sufficient energetic support.

Therefore, the answer to the question, why the universe observed today by us is so old and so big, can only be: because otherwise mankind would not be here. This leads to the Weak Anthropic Principle. It says:

> The physical universe observed by us has a structure that permits the existence
> of observers.

But is this really a principle? Principles are introduced to explain something, for
example:

- Why is the universe structured the way it is structured?
- Why is it such that life could have developed?

The Weak Anthropic Principle does not possess this explanatory merit. Perhaps
it is tautological, something like: a universe allowing for an observer to exist. Well,
what else should there be to observe? No. The Weak Anthropic Principle directs our
attention to the fact that the possibility of life is linked closely to the overall cosmic
development. Does this mean that man has moved to the centre stage again? Again:
no. The Weak Anthropic Principle only reminds us to involve the observer in any
theory to be constructed.

Another false conclusion from this principle is the following: the basic characteristics of the universe must be such that observers can develop. This is neither logical
nor required by natural laws. The only thing that matters is that, if the universe is
observed, then its basic characteristics must be such that an observer is possible.

Now there are a number of counter arguments. The first is: the denial that the
fine adjustment is essential for the existence of life, because life could be based on
something else than carbon. This means that a different type of life could have developed in a universe with different characteristics. Of course this is highly speculative.
There is no empirical evidence for a different chemistry of life.

Another objection reads like this: the existence of the fine adjustment is inevitable.
Justification: there are infinitely many cosmoses in a multiversum. In all of these, all
possible laws, constants, boundary and initial conditions are realised. Our cosmos
is then by default a necessity, and there is nothing to wonder about and to explain.
The remaining question is: what should these multi-world scenarios look like? Two
models are on offer:

- The inflationary model with many parallel cosmoses.
- The oscillating model.

One turns to the idea of a multiverse, which probably can never be verified, to
explain certain cosmological characteristics that do not conform to the traditional
standard model: anisotropic structure, missing (dark) matter, initial phase etc. The
most extensive variant of this theory postulates an infinite number of universes of all
imaginable occurrences, such that our universe also must inevitably have come into
existence as it is and as we find it. Indeed, our universe has to appear infinitely often
with all persons in it!

Or: the fine adjustment is only a clue to some still unknown context of natural
laws. The accidental must be eliminated by finding new and deeper principles.

And in the end: the fine adjustment is just random. But the improbable also does
happen occasionally.

To continue, and how should it be otherwise, there exists the Strong Anthropic Principle. It says:

> The universe has to be conditioned with regard to its laws and composition in such a way that it will eventually generate an observer without fail.

This extended formulation is logically possible but does not follow from the Weak Anthropic Principle and is basically teleological, i.e. purpose and target oriented. The origination of life is declared as a necessary characteristic of the universe. And in this way, the subject to be explained is simple postulated to be. By doing so, the need for justification is removed. The counter question is: why should the Strong Anthropic Principle be valid? Until now, objectives have always been replaced by mechanisms. Furthermore, teleological statements are difficult to falsify.

And finally, there is the Final Anthropic Principle. It says:

> In the universe, intelligent information processing life must come into being, evolve and continue to exist forever.

To conclude, one can say that, although these theses do not claim to have a specific physical explanatory value, they have quite impressively worked out that the universe is de facto arranged in a way that suffices for the origination of intelligent beings.

None of these principles, however, proposes an explanation of the way in which observing life may be generated—a life with an exposed position in the cosmos. And on the other hand, there is no conclusion as to whether this intelligence is sufficiently equipped to be able to carry out objective observations at all. The whole question concerning the Anthropic Principle is based on the assumption that our observations are indeed sufficiently correct that the Anthropic Principle makes sense at all.

Quantum mechanics has opened up the general problem of measurement, i.e. observation. It is now universally accepted that the observer is always part of the observed. From this follows a number of epistemological intricacies not to be discussed any further here, but quantum mechanics puts the validity of the Anthropic Principle into yet another perspective.

This Book

So, one of these observers was indeed Galileo. He was in the thick of the transition from the Middle Ages to modern times. This book, although it means to be more than just a repetition of time line events, still follows this sequence from cradle to grave. Most chapters, however, do not start with an anecdote from the life of our hero but introduce a specific aspect of modern science that is either directly connected to Galileo's discoveries and research results or is the consequence of further developments in relation to his findings or theories since. Through these associations, the relative position of Galileo with respect to other scientists in history can be ascertained and his contribution assessed. Chapter 7 documents the forceful clash of the world models that was at stake. This clash of ideas culminated finally in and was encapsulated by Galileo's trial and its outcome.

Chapter 2
Time and Space (1564–1642)

Time and space are fundamental conceptual constructs indispensable for the mathematical formulation of physical models of our world as we experience it with our senses. But time and space are also everyday concepts in the minds of ordinary people, necessary to accumulate their experiences and consolidate their memories. In this sense, time and space will be considered along the world line of Galileo's life whenever certain events have an impact on his professional career. Time and space in a narrower sense of course contain the personal world line of Galileo himself and thus touch upon the interest of any biographer. The space-time of Galileo's life is filled with the history of his time in and around the places he lived, in Italy, in Europe and in regions further away. These events and circumstances of the time of the vanishing Middle Ages and the dawn of Modern Times had a profound influence on his thinking and thus on the development of his contributions to science, just as they had on the reasoning and beliefs of his contemporaries, and just as our times have on us. Since however, we are concerned with the life and scientific contributions of this man, we will use this chapter to outline the context in which he lived (Fig. 2.1).

Figure 2.1 illustrates this historical context graphically.

The Time of Witch Hunts

There have always been instruments and methods for bringing people deviating from a presumed, generally accepted normality back into line. The first step is to identify these assumed deviations. At various times, the idea of witchcraft emerged already in ancient times. It was developed further and systematised during the Middle Ages to become a virulent source of social control. Witch trials appeared in waves with intermediate cessations in between (Fig. 2.2). There were peaks in the 15th and 16th centuries and again during the time span of our present concern—during the lifetime of Galileo. The emphasis shifted from the accusation of simply practising witchcraft as such by supernatural means to the accusation of direct association with the devil himself.

© Springer International Publishing AG, part of Springer Nature 2018
W. W. Osterhage, *Galileo Galilei*, Springer Biographies,
https://doi.org/10.1007/978-3-319-91779-5_2

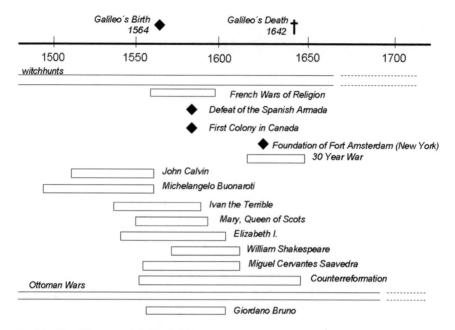

Fig. 2.1 Time Line around Galileo's Life

Witches and the devil were associated with excessive orgies, during which Satan was worshipped in the place of God. The bulk of the witch-hunts took place in Germany, although there they had begun later than in other European countries. Altogether, witch hunts lasted for approximately 300 years all over Europe. During this period, about ten thousand witches, male and female, were tried in southern Europe, including Italy, and thousands of these actually were executed, usually through burning at the stake. Galileo was spared such a trial, because he was identified as belonging to another class of deviants: heretics. The procedures to deal with these deviants were different.

The French Wars of Religion

Other rather turbulent events around Europe at the time in question, although geographically not occurring in the locations in which Galileo lived, were part of what are called the French Wars of Religion. These historic events refer to a series of eight different and successive wars over a time span of 36 years, beginning just two years before Galileo's birth. The belligerents were Protestants on one side, represented by the Huguenots, with England and Scotland and Catholics on the other side, represented by the Catholic League, Spain and the Duchy of Savoy. The wars were interspersed by various attempts at peace, such as the Peace of Longjumeau (1568),

Fig. 2.2 Interrogation of a Witch (Hans Ueli (1577): "Torture of the Wife and Daughter of a Wagoner in Mellingen"

the Peace of Saint-Germain-on-Laye (1570), the Edict of Boulogne (1573), the Edict of Beaulieu (1576), the Treaty of Bergerac (1577), the Treaty of Fleix (1580), the Peace of Vervins and the Edict of Nantes (1598), the latter of which brought some sort of end to the fighting, which had included some factional strife between different opponents within France itself. One of the unfortunate climaxes of these conflicts was the massacre on St. Bartholomew's Day in 1572. Although Galileo's life was not directly affected by these acts of war, Rome and the Pope obviously had an interest in the final outcome. The religious and political climate all over Europe and thus in Italy as well was strongly influenced by these events and shaped the attitudes of the personalities who were to deal with Galileo in the critical phases of his life.

The Spanish Armada

The French Wars of Religion were by no means the only wars that took place during Galileo's lifetime. When he was 24 years old, in the year 1588, a most historic event occurred, which is still immediately recognizable to this day with just the two words:

the "Spanish Armada" or rather: the "Defeat of the Spanish Armada". This battle was part of the undeclared Anglo-Spanish war, which raged from 1585 to 1604. The belligerent parties involved were the Kingdom of England and the Dutch Republic on the one side and the Iberian Union with the House of Habsburg on the other.

The purported reason for this conflict was again found in religious quarrels. But in fact, it was triggered by the wish of Henry VIII, who wanted to divorce his first wife, Catherine of Aragon. The divorce, however, was never granted by the Pope, and Henry VIII decided to break with the Catholic Church. After some years, the English Reformation fell into line with the Reformation in other parts of Europe. This was followed by various attempts at Counter-Reformation, but it was Elizabeth I who in the end firmly established Protestantism in England. King Phillip II of Spain, with the support of Pope Sixtus V, subsequently organised a crusade to win back England for Catholicism, and to this end he assembled the Spanish Armada to invade England. As is well known, this attack resulted in complete failure. Through a combination of battle engagements with the English fleet, commanded by Sir Francis Drake, navigational errors and foul weather the Spanish Armada was thoroughly defeated. Out of 130 Spanish ships, 35 were lost and about 20 000 men died.

The Colonization of Canada

For most scholars the beginning of the Modern Age coincides with the discovery of America in the year 1492—well before Galileo's birth. New horizons had opened up far beyond the geographical limits of Europe. This led to an enlargement of intellectual perspective and had far-reaching effects on human thought and reasoning about the world around. During the lifetime of Galileo, colonization was already well on its way—apart from all other places, notably in Canada, where it commenced with the founding of St. Johns in Newfoundland by Sir Humphrey Gilbert. It was the first English Colony on the American Continent. This was later followed by the settlements at Port Royal and Quebec at the beginning of the 17th century. New France, as Canada was initially known, was rapidly colonized along the St. Lawrence River and the Great Lakes by Missionaries and Trappers.

New York

Other renowned settlement activities started in the year 1609 further south: the Dutch began under Captain Hudson to explore the area around what is known today as New York Bay. The impulse for this effort was purely commercial and stood in connection with the profitable beaver fur trade. A number of different expeditions followed each other, resulting in the foundation of the New Netherlands in 1614. In the same year, the first European traders arrived in Manhattan. In 1624 the first permanent settlers embarked with their families, and one year later Fort Amsterdam was founded and the area around it was bought from a local Indian tribe in exchange for European goods representing a value of about 24 $ at that time—corresponding to 1100 $ in today's money—by Peter Minuit, the director of the Dutch West India Company in New Amsterdam, today New York.

The Thirty Years War

One of the most devastating and dominating catastrophes in Europe during Galileo's lifetime was of course the Thirty Years War, which was to continue for another six years after Galileo died in 1642. This war again was supposedly a war of religion—in many ways the final one in a long series of preceding conflicts. All of the parties in science, politics and the church, who will play roles in the disputes discussed later on in this book, were influenced one way or another by the developments of this conflict. Nothing in the lives of ordinary people and those in power was left untouched by this ongoing affair: economics, personal power, beliefs, employment and how to reflect on the world and its destiny. Concerning Europe, the destruction caused by the Thirty Years War was only surpassed several hundred years later by the two World Wars in the first half of the 20th century—judged by the number of direct and indirect casualties from sword, famine and disease.

Initially the conflict started between Protestant and Catholic states and alliances, triggered by the religious policies of the devout Holy Roman Emperor Ferdinand II, but it soon resolved to an already existing France-Habsburg-rivalry, involving several mercenary armies all over Europe, with Catholic France in the end even joining the Protestant Union in its fight against the Catholic League. The war erupted full scale after the Protestant King of Sweden, Gustav Adolph, came to the rescue of the faltering Protestants in Germany. The war finally ended with the Westphalia peace agreements in Muenster and Osnabrueck, which changed the political landscape in Europe entirely in favour of France and Sweden.

John Calvin

Galileo enjoyed a number of famous and illustrious contemporaries whose cultural influence is still recognizable to this day. Among them was John Calvin, the French theologian and reformer, the founder of what is called Calvinism. Calvin was born in 1509 and died in the year of Galileo's birth, in 1564. Although the full impact of Calvin's role in the changing world was not personally experienced by Galileo, the influence of Calvin's doctrines, especially the one on predestination, helped shape theological discussions during Galileo's lifetime.

Michelangelo Buonarroti

Another famous Italian who left the world in the year of Galileo's birth, but who left a legacy well beyond his lifetime and whose works could possibly have been admired by our scientist, was Michelangelo Buonarroti. He was born in 1475. As a Renaissance painter, sculptor and engineer, his influence on Western art and architecture was only comparable to that of his contemporary, Leonardo da Vinci. Already while he was still alive, he was acclaimed as the greatest living artist on Earth. Among his best known achievements figure the sculptures Pieta and David, which he created when he was not even 30 years old, and of course the ceiling paintings in the Sistine Chapel. He also took over as architect of St. Peters' Basilica, parts of which were finished under his directorship, others after his death, but still according to his original designs.

Ivan the Terrible

When Galileo became 20 years old, far away in what is now Russia Ivan the Terrible died. He had been Tsar from 1547 until then. His main achievement was the foundation of an empire uniting such regions as Kazakhstan, Astrakhan and parts of Siberia, thereby managing the transition from a small medieval state to a true empire at the threshold of the Modern Age. Some of his cultural accomplishments, like the Moscow Print Yard, and his popularity among the common people left a positive memory in the minds of some of his subjects. But he appeared to have suffered from paranoia, which seems to have been the cause of his treatment of his perceived enemies with the utmost cruelty.

Mary Stuart, Queen of Scots

Mary Stuart, known as the Queen of Scots, was born in 1542 and was beheaded in 1587, when Galileo was 23 years old. Her reign in Scotland lasted from 1542 (when she was six days old) to 1567. During this period, she lived in France for most of the time, and Scotland was administered by her proxies. After marrying the Dauphin of France in 1559, she also became briefly Queen of France until the death of her husband in 1560. Back in Scotland, she was forced to abdicate several years later and fled to England. Instead of giving her shelter, Queen Elizabeth I was ultimately instrumental in the execution of her rival to the throne. At the same time, the conflict between the two cousins was mirrored by the escalation of a Catholic uprising to undo the Reformation in England.

Elizabeth I

We have already touched upon events of the Elizabethan era, the period from 1558 until her death in 1603 during which Elizabeth I ruled England. She was born in 1533 as the daughter of Henry VIII and Anne Boleyn, who was executed when Elisabeth was two years old. It was Elizabeth, after her succession to the throne, who founded the Protestant Church of England.

William Shakespeare

One of the most prominent figures of the Elizabethan era was a poet and playwright, born in the same year as Galileo, 1564, in Stratford-upon-Avon: William Shakespeare. He died in 1616. His 38 plays are the most performed plays worldwide to this day. He married early at the age of 18 and had three children. He started his career as an actor, but soon wrote his own pieces for his own company of actors. His repertoire included comedies, historical dramas and tragedies. His works were published and distributed widely already during his lifetime.

Miguel Cervantes Saavedra

Another famous writer of that time was of Spanish origin: Miguel des Cervantes Saavedra, author of Don Quixote, born in 1547. Part of his early life he spent in Rome in exile, between 1569 and 1571, thereafter, between 1575 and 1580, in the captivity of Ottoman pirates, who captured him during operations in the Mediterranean, after Cervantes had joined the Spanish navy. Don Quixote was published to immediate success in 1605. Cervantes died in Madrid in 1616.

The Counter-Reformation

While Copernicus at his time enjoyed a more or less peaceful existence—a servant of the church until his death—times had changed considerably, when Galileo encountered his theory of celestial mechanics as presented in "De Revolutionibus Orbium Coelestium" half a century later. The gathering momentum of the Counter-Reformation produced a situation far more dangerous for people who were considered heretics than it had been earlier. The Counter-Reformation, intended to undo the Protestant Reformation, started off with the Council of Trent, which lasted from 1545 to 1563, the year before Galileo was born, and ended effectively only with the end of the Thirty Years War. The Counter-Reformation was not only a constructive response to Protestant criticism of past excesses of the old church, including structural reforms and a new spirituality, but was also highly political, striving for power in Europe and in the new colonies. One of its most effective instruments was the Holy Inquisition, at the hands of which Galileo had to suffer, when his time was due.

The Ottoman Wars

Other military developments contributing to the political climate of those times were the Ottoman Wars, which started in the late Middle Ages and created a sense of ultimate existential threat both in the ruling circles and in the general population (the Turkish Wars continued well into the early 20th century). These wars were conflicts between the Ottoman Empire and various individual European states—taking place in the Balkans and progressing well into Central Europe with the siege of Vienna in 1529 as a high water mark 35 years before Galileo's birth. During his lifetime, the European central powers tried to reverse the gains of the Ottomans and to reclaim lost territory.

Giordano Bruno

One other prominent contemporary of Galileo, who fell victim to the new rigidities of the Holy Inquisition, was Giordano Bruno (Fig. 2.3), born in the year 1548. His views extended far beyond the Copernican model of the universe. In fact, Bruno proposed that the distant stars were of the same nature as our sun. His model of the universe was a very modern one, already containing elements, some of which were developed and refined only in the 20th century. Bruno believed—although he could not prove it at that time—that the universe was infinite and had no centre. This latter insight was accepted only in our times as a consequence of the General Theory of Relativity. Bruno was tried for heresy and burned at the stake in 1600, when Galileo was 36 years old.

Italy

Italy was at that time by no means a nation-state such as France or England. Even before Galileo's birth, the country was dominated by foreign powers, notably Spain under Habsburg's emperor, and internal strife was rampant. The European major powers, among them the rising France, were engaged in a bitter political fight for influence. The Pope and the Catholic Church relied mainly on the feudal system to maintain its own power politics, were intent on remaining independent from other major rival Catholic powers, and were thus inclined to rather mutable support of one

Fig. 2.3 Giordano Bruno

or the other of them, the allegiance changing with the situation. Living conditions had been degraded. The discovery of America lead to the decline of the importance of Italian ports against those of Spain or Portugal, which had access to the Atlantic. In 1630, the plague devastated the country, claiming the lives of nearly 25% of the population. Since the second half of the 16th century, the bourgeoisie tended to invest in rural property rather than in other businesses since these had stagnated after trade routes had lost their pre-eminence.

It was definitely a time of change. Ordinary people, scholars and the powerful in state and church had to come to terms with a world, which—during a relatively brief span of time—had passed through major turmoil and presented entirely new perspectives for an unpredictable future to them. Nothing would be the same again, but—as is always the case in any other historic turnover—most persons still tended to hang on to what had been their spiritually secure past. This was the time into which Galileo was born and in which he lived. His personality was still very much dominated by the spirit of the Renaissance. However, he had the choice either to stand against the rigours of the Counter-Reformation dogmas by relying on his own reason or to continue to subordinate himself to it.

Chapter 3
Early Years (1564–1588)

On July 4th 2012, the CERN Laboratory in Geneva announced the discovery of a new scalar particle with a mass of about 126 GeV. This mass corresponded to the one predicted by the so-called Higgs mechanism. All other properties of this particle fitted the predictions of this theory as well. What is behind the Higgs theory?

In the sixties of the past century, the physicist Peter Higgs and other researchers had contemplated a fundamental problem in connection with gravitation. In analogy to the electrical charge, they proposed that one could define a gravitational "charge", presented by the mass of a body. So how does an elementary particle or any other body obtain its mass in the end? Higgs and his colleagues theorized a field, later called the "Higgs field", which is omnipresent, interacts with all particles and thus bestows them with their respective masses. The exchange particles of this field are called Higgs Bosons. The discovery announced by CERN was in fact that of the Higgs Boson.

The hunt for this particle was one of the justifications for building the enormous and costly configuration called the LHC (Large Hadron Collider) at CERN to enable proton collisions at the high energies required to produce this boson. This apparatus represents the culminating step in the research and theoretical efforts concerning the weight, the mass and the movements of bodies in nature, dating from Aristotle and Archimedes, via Galileo and Newton (Fig. 3.1).

Archimedes

In fact, the first persons who occupied themselves with the secrets of mass and weight belonged to those Greek Pre-Socratic philosophers called the "Atomists", particularly Leucippus and Democritus. They developed the first theoretical thoughts about momentum and the attraction of masses.

However, the most important contribution in ancient times regarding the question of weight came from Archimedes. Archimedes was born as the son of the astronomer Phidias in the port town of Syracuse in Sicily around the year 287 BC. He died in

© Springer International Publishing AG, part of Springer Nature 2018
W. W. Osterhage, *Galileo Galilei*, Springer Biographies,
https://doi.org/10.1007/978-3-319-91779-5_3

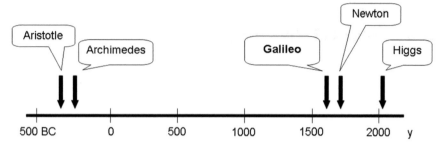

Fig. 3.1 The history of weight and mass

the same place in 212 BC, the year in which his hometown was conquered by the Romans during the Second Punic War. Archimedes became famous because of his many mechanical inventions and his substantial contributions to the mathematics of his time.

The single discovery most relevant to our subject was a principle, later named after him, the Archimedean Principle. According to legend, he was asked by the ruler Heron II to determine the content of gold in a crown dedicated to the Gods. The ruler was suspicious that the goldsmith had cheated him with respect to the amount of gold employed. Archimedes prepared a bar of gold with the same weight as that of the crown and submerged successively both objects in a vessel filled with water, then measured the amount of water displaced by each object by capturing the water overflow. The result was that the crown displaced more water than the test bar. The philosopher concluded that the specific weight of the crown therefore was less that that of the gold bar, thus also containing other lighter materials than gold. The important discovery was that of hydrostatic uplift and the specific weights of objects of different materials. Galileo later doubted the legend and believed that Archimedes must have employed a balance to measure the buoyancy of the bodies.

Aristotle

The dominating figure influencing natural philosophy and thus physics up to the end of the Middle Ages and even well into the Modern Age was the Greek philosopher Aristotle. Aristotle was born in 384 BC in Stagira and died in 322 BC in Chalcis. His subjects of interest comprised natural philosophy, logic, biology, physics, ethics, state theory und the theory of poetry. Relevant to our considerations here are his views on natural motion.

He founded his interpretation of free falling objects on the assumption that all bodies were composed of a mixture of the four elements earth, water, air and fire. The ratio of these elements constituting a physical object would determine its hierarchical position in the cosmos. Every object had its place to which it would eventually want to migrate after having been displaced by outside forces beforehand. And all objects had a tendency to migrate to the centre of the universe, in which our Earth resided. These were the boundary conditions determining the speed with which a specific body would fall in a particular medium. Motion was thus a quality of the object in question itself.

This interpretation was still predominant during Galileo's time. We shall come back to this controversy later, when we discuss Galileo's free fall experiments (Chap. 4) and his most controversial major work, the "Dialog about the Two Major World Systems" (Chap. 8). For the time being, it is important to note the influence of Aristotle's thoughts on the interpretation of weight.

Isaac Newton

426 years had to pass between Galileo's "Bilancetta", the first publication in which he concerned himself with weight, and the discovery of the Higgs Particle. And the one important contribution on the natural force of gravitation in the intervening time was to be found in the work of Isaac Newton. Generally speaking, up until Newton, physical sciences were concerned with the more descriptive aspects of matter and its motion. This was particularly so for astronomy. The example of Kepler, which will be discussed in Chap. 7, shows very clearly that the emphasis even at that time still was on cosmological harmony. Mystical sources continued to determine the thinking of even the most brilliant minds of science at the dawn of the Modern Age.

The world was perceived as it was. The reasons why it was as it was were not contemplated at all for a long time. The movements of celestial bodies could be described with some accuracy, but the explanation for their movements had to wait for the genius of Isaac Newton.

Newton was born in 1642, the year in which the civil war between the crown and parliament broke out in his country. In the year of his death, 1726, Jonathan Swift published his novel "Gulliver's Travels". His work most relevant to this chapter was "De Motu Corporum" ("About the Movements of Bodies"), published in 1684 (we will come back to other relevant aspects of Newton's findings later in this book). In this publication, Newton summarized his own mechanical experiments. His practical and theoretical results were later incorporated in "Philosophiae Naturalis Principia Mathematica" ("Mathematical Principles of Natural Philosophy"). In this major work, he consolidated the results of motion experiments by Galileo, Kepler´s observation of the movement of planets and reflexions by Descartes about inertia. His three laws of gravitation were the foundations of classical mechanics [2]:

1. "A body remains in a state of rest or uniform straight motion, if it is not acted upon by interacting forces to change its state."
2. "A change in a movement is proportional the acting force and proceeds along the straight line, on which the force acts."
3. "Action is equal to reaction; or the actions of two bodies upon each other are always equal and in opposite directions."

The second law is most relevant to us at this point, since the factor of proportionality is the mass of a body. This brings us now to Galileo himself.

From Pisa to Florence and Back

Galileo Galilei was born on February 15th 1564, a Tuesday, in Pisa (Fig. 3.2) at half past ten at night. His baptism took place four days later on the 19th. His parents, living in the countryside near Pisa, were impoverished Florentine patricians. His father Vincenzo Galilei originally came from Florence, was a cloth trader, but had a variety of other interests, ranging from music to mathematics. The name of his mother was Giulia degli Ammannati, from Pescia. Little else is known about her origins. Galileo was their first child. He was joined later by five more brothers and sisters.

At the time of his birth, the intellectual climate in his home country was widely influenced by the writings of Mirandola about human dignity, of Lorenzo Valla about the free will and particularly by Machiavelli's ideas, trying to distinguish between the realm of the divine and human endeavours. The thoughts of Machiavelli would later encourage Galileo's positioning with respect to the predominance of reason and his attempts to assert it.

In 1572, when Galileo was eight years old, the family relocated to Florence. They left their eldest son behind with a close relative, Muzio Tedaldi, for another two years. After this, he re-joined his family in Florence, where he was tutored by Jacopo Borghini. Already at that time, at this early stage, his interest in mathematics and physics had been awoken by the pastimes of his father.

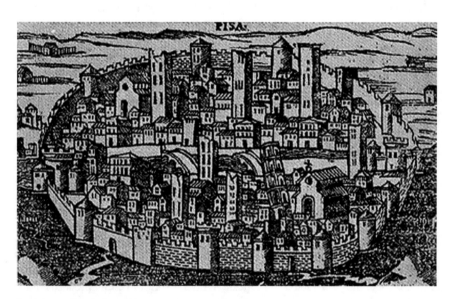

Fig. 3.2 Medieval view of Pisa

When he was fourteen years old, Galileo was accommodated in the monastery
Maria di Vallombrosa near Florence, run by Vallombrosian monks, an offshoot of the
Benedictines, as a novice. He developed a strong inclination to join the Benedictine
Order. But in 1580, he was called back to Pisa, where he lived once more with his
relative Tedaldi, to study medicine at the University of Pisa. This was the wish of his
father. He stayed enrolled at the Faculty of Arts from September 5th 1580 onwards
for altogether four years. But from the very beginning of his studies, the subject of
medicine never really caught his interest. He preferred rather to join courses given
by Ostilio Ricci, a scientist from the school of Nicolo Tartaglia, a mathematician
famous for his solutions of cubic equations, to attend his lectures on Euclidean
Mathematics—all 15 books of the "Elements", including basic geometry, theorems
of triangles, planes and circles. Ricci was at the same time mathematician to the court
of Tuscany for Francesco de Medici (Fig. 3.3).

During his enrolment at the university, Galileo encountered the writings of Aristotle. These were compulsory subjects, including "Physica", "De anima", "De caelo".
These were one important source of Galileo's understanding of the world. The other
impulse came from the observation of nature itself.

As Galileo's interest in medicine was on the wane and his interest in mathematics
on the rise, a conflict with his father became inevitable. When Galileo returned home
for his summer recess to Florence and told of his inclinations, his father strongly

objected. Galileo even invited Ricci to persuade his father that his main interests were really in mathematics and in nothing else. Vincenzo Galilei continued his resistance. Eventually a compromise was reached. His son was allowed to study the works of ancient mathematicians like Euclid and Archimedes, but remained enrolled in Pisa for medicine for the time being. In the end, Galileo gave up in Pisa in 1585 without having obtained a degree.

Back in Florence, he continued to attend the lectures of Ricci. He earned his living partly by giving private lessons to other students, later, from 1585 to 1586, in Siena and also at Vallombrosa, his old place of instruction. His subjects included applied mathematics, mechanics and hydraulics. Right at the start of his career he—like other contemporaries as well—was confronted with the challenge to cross thresholds established far in the past and taken for granted ever since. These thresholds circumscribed and protected the visions and models of the world as it was supposed to be.

In 1587, he turned to the problem of weight for the first time, when he produced a paper called "Theoremata circa centrum gravitates solidorum" ("Theory about the Centre of Gravity of Solid Bodies"). He sent copies of this work to a number of famous mathematicians of that time, among others to Guidobaldo del Monte in Rome, a teacher at the Jesuit College of Higher Learning, whom he later visited personally to discuss his findings with him. Its first official publication, however, had to wait for another fifty years. Galileo then took up the subject of gravitation later again in his famous Dialog (Chap. 8).

He based his first essay on the writings of Archimedes about the centre of gravity of plane surfaces. Galileo proposed a series of thought experiments involving scales, on which a number of different weights were attached at different points away from the geometrical centre of the scales. He then calculated the centre of gravity for every configuration. He found that for each setup the ratio of the weights on both sides of the centre corresponded to the ratio of the geometric divisions of the scales themselves. This approach was somewhat in contradiction to Aristotelian teaching, generally accepted at that time, that the weight of a body was a quality derived from its position in the cosmos due to a natural order: heavy bodies should be positioned nearer to the centre of the Earth, the centre of the cosmos, lighter bodies further away. According to this theory, a heavy body should tend to fall towards the Earth and lighter materials like smoke would tend to rise up. We shall come back to this controversy between experimental observations and Aristotelian philosophy later in the chapter about the free fall (Chap. 4).

La Bilancetta

Galileo began to make a name for himself in Florence by giving a lecture about "The Topography of Dante's Hell" at the Accademia Fiorentina in 1588. At about the same time he published his treatise "La Bilancetta". Figure 3.4 shows the title page of the edition of 1656 after Galileo's death.

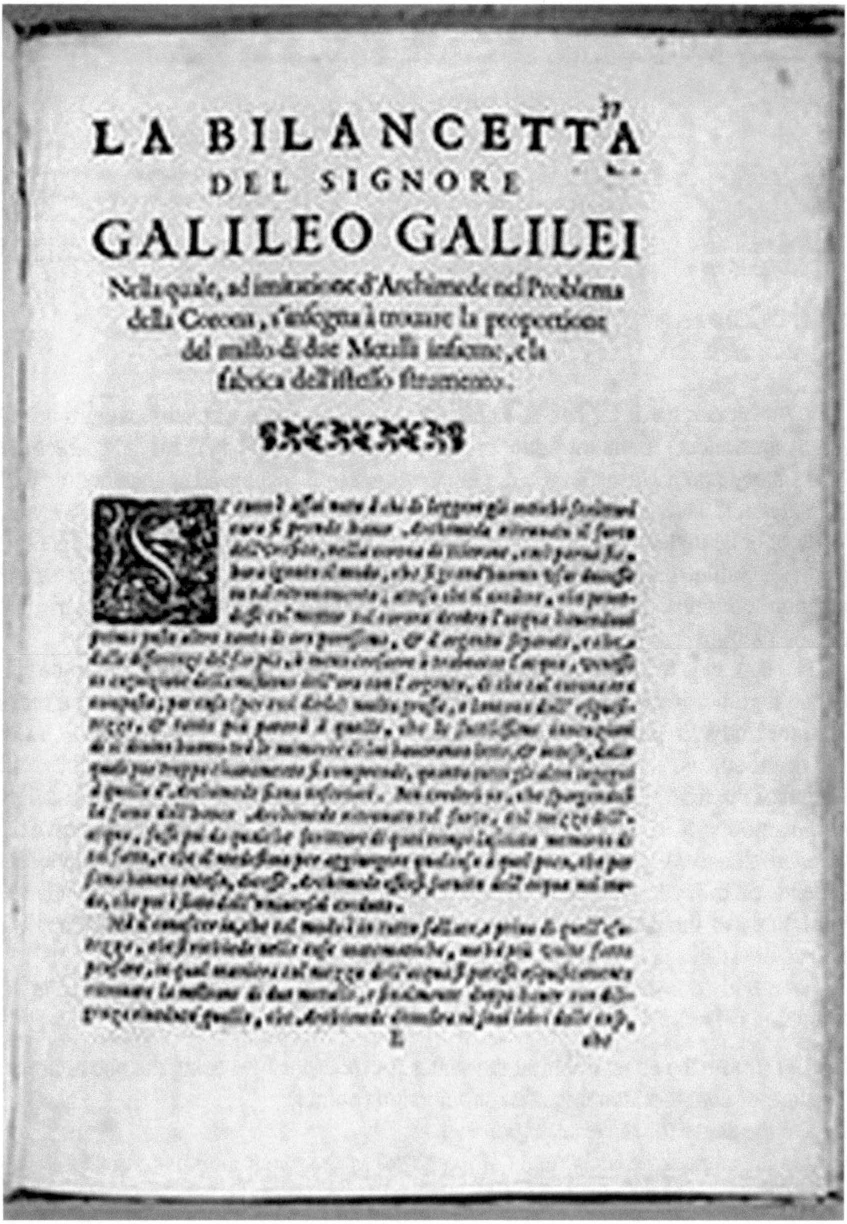
LA BILANCETTA
DEL SIGNORE
GALILEO GALILEI
Nella quale, ad imitatione d'Archimede nel Problema della Corona, s'insegna à trouare la proportione del misto di due Metalli insieme, e la fabrica dell'istesso strumento.

Fig. 3.4 Title Page of "La Bilancetta" Bologna 1656

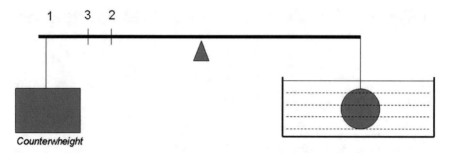

Fig. 3.5 La Bilancetta

In this paper, he described an approach for measuring the specific weight of different substances. In the introduction, he expressed his doubts about the correctness of the story about Heron's crown. Galileo suggested that some ancient writer with no real mathematical understanding had heard this story and then written down what he thought he understood. In fact, after having studied Archimedes' "About Floating Bodies", Galileo believed that the ancient Greek must have constructed an instrument far more complex than the story implied. So he himself now developed a method that he thought Archimedes might have used.

The task was to find a device to determine the proportion of different materials mixed together in one solid body. Galileo proceeded in several steps, using a set of scales. Firstly, he proposed to place a solid body made of a single material on one end of the scales. This should be counterbalanced by another body having the exact weight of the first. If the first is now submerged in water, it weighs less by the weight of the amount of water it displaces. To balance the scales, one now has to move the counterbalance weight closer to the centre of the scales. This adjustment obviously depends on the specific weight of the solid body that is made of one single material. For a body of the same size but made of lighter material, the adjustment along the arm of the scales is less, and more, if the material is heavier.

For a body consisting of a mixture of two different materials the sequence is the following (Fig. 3.5):

1. Determine the exact point on the scales for material 1 by using the above procedure of counterbalancing after submersion (point1).
2. Do the same for material 2 (point 2).
3. Do the same for a body made of a mixture of the two materials (point 3).

The point of balance for the latter must be somewhat between the one for the heavier and the one for the lighter material on the scales. The ratio of the composition of the two materials corresponds to the ratio of the distances between point 1 to point 3 and point 2 to point 3.

In "La Bilancetta" the 22-year-old Galileo exhibited for the first time his unique capability of combining mathematical reasoning with technical craftsmanship. He also broke with the established tradition of publishing scientific works in Latin, but instead used the everyday language "Volgare", much as Luther had popularised the German language in his theological writings earlier on.

Chapter 4
Pisa—Gravitational Experiments (1589–1592)

ZARM

Cuxhaven is a popular holiday resort on the North Sea coast about 100 km northeast of Hamburg. People arriving from southern Germany usually take the Autobahn A1 until Bremen and then turn onto the A59 in the direction of Cuxhaven. Shortly after making the turn, they will notice a very tall, slim building on the left hand side—the drop tower of ZARM. ZARM stands for: "Zentrum für angewandte Raumfahrttechnologie und Mikrogravitation" or "Centre of Applied Space Technology and Microgravity". It was founded in 1985. Its main facility is the Bremen Drop Tower. The first question that comes to mind is: why does somebody need another drop tower? Did not Galileo exhaustively use the Leaning Tower of Pisa to investigate falling objects under the force of gravity?

Firstly, there are serious doubts about Galileo dropping objects from the Leaning Tower at all, and secondly the purpose of ZARM is not to repeat Galileo's experiments some 400 years later to measure the acceleration in the gravitational field of the Earth with better precision. Its main research is—as its name says—the investigation of micro gravitational effects. Researchers working in this facility prepare for example microgravity experiments for the International Space Station (ISS). This is a long way from Galileo's observations.

Eoetvoes and Others

But still, even 400 years after Galileo and some 200 years after Newton, researchers were not entirely satisfied with the results of these earlier free fall observations—the result having been the exact value of the acceleration in the gravitational field of the Earth ($g = 9.81$ m/s^2). In 1906, the Hungarian physicist Lorand Eoetvoes objected to the common assumption that a free falling body would do so in a purely vertical direction. He claimed that there must be a force component in the direction of the

© Springer International Publishing AG, part of Springer Nature 2018
W. W. Osterhage, *Galileo Galilei*, Springer Biographies,
https://doi.org/10.1007/978-3-319-91779-5_4

Earth's rotation as well. In his Budapest laboratory, he used a special torsion balance to measure it. He determined the maximum value for the deviation from g was less than 10^{-9} from the established value.

Nevertheless, in 1964 the old controversy, whether the acceleration of an object depended on the type of material it is made of or not—long thought to be resolved by Galileo, Newton and others—was again revived by three American researchers, Roll, Krotkov and Dicke. They developed the Eoetvoes experiment further, including a variation factor due to the attraction of the sun. The measured difference between two bodies, one made of gold, the other of aluminium resulted in a maximum deviation value of less than 10^{-11}.

The Nature of Force

Of course, all this has to do with the concept of force itself. The origin of the word "force" appears to be mystical. Although today, people believe that they use it in a rational way, it was coined in pre-scientific times. "Force" still stands for something mysterious, quite often something menacing from a remote distance, emanating from powers that somehow have to be appeased. In latter times, scientists have appropriated this term from the irrational realm to describe the observable world with its help, using it in formulas, and thus have made it into one of the most important fundamentals in physics. By employing this construct, they surely created one source for the complexity of the resulting mathematical body. Later, Einstein tried to get rid of the "force" once and for all by replacing it with geometrical structures. However, since human reasoning at his time had long been quite familiar with the old mystical concept, his approach caused even more headaches for most people.

To sum it up:

> The environment, in which a force acts, is called a field of force. The first classical field of force that was described qualitatively and quantitatively is the field in which the gravitational force acts: Gravitation.

Forces have impacts. The remote impact has already been mentioned. Under gravitation, masses in the form of enclosed bodies separated from one another interact in fact with one another across the intervening distance. In the case of gravity, they interact by attraction. In electromagnetism, the interaction may also be to repel each other.

As already mentioned, Albert Einstein followed an entirely different approach to deal with the gravitational force: he wanted to get rid of it. His claim was that physicists had all along used the wrong coordinate system, which must lead to the classical relationship between force and acceleration. If only one would employ "natural" coordinates, acceleration would disappear. The consequence was that instead of

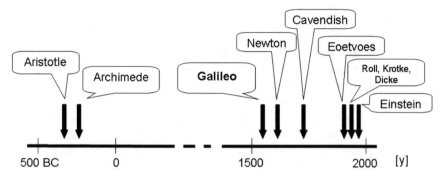

Fig. 4.1 The history of force

mass being a proportional factor between force and acceleration, mass would cause the curvature of space around it, the space in which an object moves, thus forcing the object onto a trajectory within the curved space that is very similar to the one displayed by an acceleration graph in classical Cartesian coordinates.

This was the result of his General Theory of Relativity (GTR), actually an enhancement ("generalization") of his earlier Special Theory of Relativity (STR), which dealt with non-accelerating objects (having constant velocity) only. We shall come back to the transformation equations of STR later.

To sum GTR up, the overall statement is:

"If one assumes that in two separate systems the same physical laws are valid, then there exists no reference system for absolute acceleration, just as there is no absolute velocity in Special Relativity."

Instead of taking space and time as separate entities, the GTR invented the concept of space-time, in which an object moves along its world line. The shortest possible distance between two events in space-time is called a geodesic. The final equation of the GTR describes the equivalence between the space-time curvature tensor (the Ricci, a Riemann-like, tensor) on one side of it and the energy-momentum tensor on the other side. Because of the equivalence of energy and mass from SRT, mass is in effect causing the curvature of space-time, leading to the movement of a body along a geodesic, just as in classical mechanics the acceleration of a body described in flat coordinates is caused by a force proportional to the mass of this body. The encoding of the classical world had started with Galileo and was later completed by Newton. Figure 4.1 illustrates the relative historical position of Galileo in the quest to understand the nature of "force".

Newton and Cavendish

Isaac Newton was born in the year of Galileo's death, in 1642. As early as 1665, he started to be interested in mechanics. That was after his return home from his studies at Trinity College. But his main fields of interest were mathematics and optics. After turning away from the natural sciences altogether for several years due to controversies with members of the Royal Society, Newton was busy with Alchemy and Patristic. But from 1679 onwards, mechanics became the centre of his attention. As already mentioned in the preceding chapter, in 1684 he published his work "De Motu Corporum"—"on the movements of bodies". This was to be the basis for his later "Philosophiae Naturalis Principia Mathematica", which contained his famous three postulates about gravitation.

He knew of Galileo's experimental results concerning free falling bodies. The legend about his observation of an apple falling from a tree and from that, deducing the gravitational force binding the moon to the Earth is as improbable a story as that about Galileo dropping objects from the Leaning Tower of Pisa. We will discuss Newton's conclusions about the movements of celestial bodies in Chap. 7.

Newton, however, made an assumption that is still as bold today as it was then: the physical laws governing events on Earth are equally and for all time past or present valid anywhere in the cosmos. Considering how small our habitat is in comparison to the rest of the world, this is indeed a rather risky assumption—but it seems to hold to today. The numerical proof for events at least on our planet was not done by Newton himself but some hundred years later by Henry Cavendish (1731–1810) [3]. In 1798, Cavendish applied a torsion balance to determine the gravitational constant. This was achieved by measuring the gravitational forces between two masses at the end of dumb bells (s. Fig. 4.2). One dumb bell was fixed, the other suspended by a wire. When released, the one suspended by the wire turned to the fixed one due to gravitational attraction. The torsion of the wire was observed by a telescope from outside the experimental room so as not to disturb the action of the instrument. This torsion finally gave a measure of gravitational attraction. Thus it was possible for the first time to determine the gravitational constant to be $k = 6.67 \times 10^{-11}$ [m^3 kg^{-1} s^{-2}].

At the University of Pisa

The University of Pisa was founded 1343, although the first scholars were present in Pisa itself already two centuries before. They had been mainly interested in the medical sciences. Formal studies commenced in 1338, initiated by Ranieri Arsini, who taught civil law. Pope Clemens VI finally recognized studies at the Pisa institution as a Studium Generale, in this way attesting it as a university. Pisa was one of the first universities in Europe at that time. Subjects taught were: theology, civil law, canonical law and medicine. It is assumed that Galileo later followed lectures about

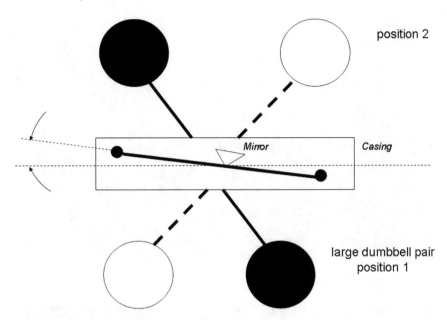

Fig. 4.2 Cavendish experimental setup (*Source* https://sciencedemonstrations.fas.harvard.edu/presentations/cavendish-experiment)

Aristotle's "De caelo", "Physica" and "De anima". These lectures were given by the professors Girolamo Borro, Francesco Buonamici and Francesco de Vieri.

When the Florentines conquered Pisa at the turn of the 15th century, they closed the institution down in 1403. It only resumed normal operation in 1473 after the intervention of Lorenzo de Medici. In 1486 a special building at the Piazza del Grano, the Palazzo della Sapienzia (palace of wisdom; Fig. 4.3), was inaugurated to house the lectures.

After the war against Florence in 1494, studies were relocated to Pistoia, Prato and Florence. In Pisa, the University was reopened in 1543, and from thereon its courses were enriched by the chairs of botany, anatomy and mathematics, the latter chair held by Galileo in 1589.

According to Viviani, his pupil and later first biographer (Fig. 4.4), Galileo obtained his post in Pisa through the good services of Guidobaldo de Marchesi dal Monte, a mathematician, to whom Galileo had demonstrated some of his earlier achievements in mechanics and geometry, among them his findings about the centre of gravity of bodies based on the writings of Archimedes. It had been dal Monte himself who previously had suggested to Galileo to tackle the problem of the centre of gravity. He was so impressed by this young man that he recommended him for the Chair of Mathematics in Pisa to Arch-Duke Ferdinand I and Prince John de Medici.

One word about Viviani's biography of Galileo: Viviani published his "historical description of the life of Galileo Galilei in 1654. In line with the customs of his time, the biography was meant not only to present the salient facts of the life of a

Fig. 4.3 Palazzo della Sapienzia (ISailko)

Fig. 4.4 Domenico
Tempesti: Vincenzo Viviani
1600

renowned personality, but also to contribute to his glory in the ensuing ages. Thus Viviani took some liberties, for example to wipe away the stain of heresy from the memory of his master, and thus contributed to some of the legends around him. The dates given by this biographer, for example, for the formulation of the laws of free fall and Galileo's alleged pendulum experiments were advanced to an earlier period of the scientist's life to underline the genius of the man. Viviani's most extraordinary attempt at glorification was the change of Galileo's date of birth to accord with the death date of Michelangelo on February 19th rather than February 15th 1564, the correct date of Galileo's birth.

In any case, the money Galileo earned in his new position was barely sufficient to provide for his living, and he had to rely on continued support from his father. On the 15th of November 1592, Galileo wrote to his father that he had received clothing and books from him and confirmed that he was still studying and following the teachings of Mazzoni. And only about three weeks later Galileo received a letter from Guiobaldo, inquiring if he had finally received a rise in his pay. In this same letter, Guiobaldo also asked whether Galileo was still working on centre-of-gravity problems, in which he still had a strong personal interest. In spite of all his financial difficulties, Galileo succeeded in constructing scientific instruments for his own use, including a simple thermometer.

Criticism of Aristotle

Galileo was not the only person of his time who doubted the postulates of Aristotle's teachings on dynamics. One well-known critic was Giovan Benedetti, who published a tract entitled the "Demonstration of proportions of local movements against Aristotle and all philosophers". Another famous critic was the mathematician Nicolo Tartaglia, who investigated free fall and the trajectories of thrown objects. The stimuli that Galileo derived from him influenced his own teaching of dynamics. Thus, the young scholar combined the knowledge of antiquity with that of scholars of the Renaissance.

After having acquired the necessary technical instruments and mathematical apparatus, Galileo felt confident enough to put Aristotle and Ptolemy to the test, which in the long run meant no less than to call the medieval world view into question. To do this, he had to abandon the scholastic methods of the past in favour of experimental ones.

At about this time, the first of many legends about this scientist was invented. One enduring legend, repeated even many years later, claimed that Galileo dropped objects from the Leaning Tower to measure the acceleration of freely falling objects. There is no mention of this in Galileo's own writings. It was Vincenzo Viviani who reported this story in his biography of Galileo. At the time in question, there existed no clocks capable of measuring the elapsed time of such a free fall. It is possible that Viviani inferred this story from a later thought experiment in Galileo's main opus, the Dialogo. However, Viviani's teacher definitely investigated the movements of a pendulum, the isochronisms of which were discovered in 1583. Galileo found that its period did not depend on its deflection nor on the weight attached, but only on its length (which in fact only holds for small deflections).

In any case, he determinedly followed his own programmatic guideline: "Ignorance of motion is ignorance of nature" and conducted his experiments to test the Aristotelian postulate that the velocity of falling bodies increases with their weight. Galileo consolidated his experimental results in a paper entitled "De motu"—about movements. The mathematical apparatus for his observations was based on experiments conducted with an inclined plane setup. He experimented with balls made from different materials, for example from lead or wood. By this means, he succeeded in measuring the speed of slowly rolling balls. This method was far more reliable than simply dropping these objects from a tower, because in this way he could minimize the effect of air drag and other influences. And thus he discovered acceleration. Although he initially did not publish his results, it became obvious to him that in any case any movement rested on measurable quantities.

This is an extract of one of Galileo's laboratory reports:

"Description of the Setup and Results:
We used a board or plank, which was about 12 cubits in length, half a cubit wide and 1 inch thick. On the narrow side, a perfectly straight groove with a width of a third of an inch was engraved. We smoothed and polished it and equipped it with parchment as smooth and as polished as possible. In this groove, we released a hardened, smooth and perfectly rounded ball of bronze. We mounted the one end 1 to 2 cubits higher as the other and let the ball role along the now askew positioned groove, as I pointed out already. We determined the time for rolling down by a method to be described elsewhere. We repeated this experiment several times to increase the exactitude of time measured in such a way that the deviations between two observations each would never be more than a tenth of a pulse rate. Once this was achieved and we were confident about the method we let the ball role down only one fourth of the total length of the groove; when we measured the time span necessary for this we realised that it corresponded to exactly one half of that measured in the first experiment. We investigated other distances and compared the time necessary for passing through the total length of the groove with that required for half, two thirds, three quarters or any other fraction of it. In these experiments, which we repeated all a hundred times we always obtained the result that the distances covered were proportional to the square of the times measured. This was the case for every inclination of the plane, i.e. the groove, over which the ball was let to role. We also observed that the run-times for different inclinations of the plane corresponded to exact the ratio inferred and forecasted by the author for them ...."

"Description of Time Measurement:
To measure the time we used a big water filled vessel placed in an elevated position: on its bottom a tube with a small diameter, through which a thin jet of water sprayed out, was soldered. During the run-time of the ball over the total length of the groove or over a fraction of its length the leaked water was collected in a small glass and subsequently weighted with the help of a very exact scales; the differences and ratios

of the weights gave us the differences and ratios of the times with such an accuracy that—in spite of many, many repetitions—no noticeable deviations of the measured values appeared" [4].

Beside his serious studies on the subjects cited so far, Galileo occupied himself in his spare time with poetry and seems to have written a number of poems himself. He was also called on to comment on contemporary poets in Tuscany, although the results of his investigations were later lost.

Unavoidably, Galileo ran into controversies with scholars adhering to traditional concepts of natural philosophy, while he himself continued to study and teach Aristotle, Plato and other ancient philosophers. The unfavourable reception of his doubts and questions made him known as a "spiritus contradictorii". In the end his contract was not prolonged after 1592, a fact that hit him all the harder, since his father had died the year before, and all sources of income seemed to have dried up.

Inertia and the Galilean Principle of Relativity

Galileo was the first person to have formulated the physical laws of inertia. This is why the transformation of inertial mechanical systems is still called "Galilean Transformation," and the associated principle of the equivalence of different reference systems is called the "Galilean Principle of Relativity". It says:

All inertial systems are equivalent. Within them the same physical laws apply.

What does this mean?
In practice, the set of mathematical equations necessary to calculate the space and time coordinates of a specific point from one inertial system to another is called Galilean Transformation. The Galilean Transformation implies that in the absence of an external force a body either remains at rest or moves in a straight line with constant velocity. Preconditions for the validity of this transformation are that the reference systems, in which these events happen, neither accelerate nor rotate.

However, in the end, the Galilean Transformation proved to be true only for systems moving much slower than the velocity of light, i.e. for classical physics. When Michelson and Morley tried to apply this transformation to the results of their experiment to measure the velocity of light with and against the movement of the Earth, the Galilean Transformation failed. The associated transformation equations predicted that the velocity of light in the direction of the movement of the Earth should differ from that against the direction of the movement of the Earth. The measured results, however, showed that this was not the case: the speed of light remained the same in all directions—independent of the speed of the light source itself.

To explain this phenomenon, Einstein enhanced the Principle of Relativity:

All physical laws are equivalent in every inertial system. Therefore, inertial systems can basically not be distinguished from one another.

The corresponding transformation in mathematical terms is called "Lorentz Transformation". It replaces the Galilean Transformation of classical mechanics in Special Relativity. The consequences resulting from this transformation are manifested in,

• Unsimultaneousness.
• Time dilation.
• Length contraction.

De Motu

Galileo called his first important work "De motu antiquiora", better known as "De motu", written between 1589 and 1592. Most scholars agree that it was written during the Pisan period. It consists of 23 chapters and a handful of additional notes. However, it was not published as such before 1854, when it was added to an edition of Galileo's works by Eugenio Alberi.

For a long time, Viviani's account was seen as one of the most important sources for understanding Galileo's work in Pisa. We now know that Viviani's treatments—as was customary und legitimate at that time—were enhanced to reflect the admiration of this biographer for his teacher. Viviani mentioned other professors at the university who took part in the discussion on the interpretation of Galileo's experimental results. In the subsequent academic dispute, Borro and Buonamici played a role.

It was by no means Galileo who first addressed the problem of the motion of bodies with different weights. The dispute had been going on in Greek, Arabic and other literature for a long time before—but mainly on a philosophical basis. Galileo's approach was experimental, and his results contradicted the philosophical consensus. But even concerning his methods, he again was not the first to apply them. There had been predecessors who had tested Aristotle and disagreed with the ancient philosopher.

In 1575 Borro published a paper called "De motu gravium et levium" [5]. He stated that the motion of each of the four elements, from which all other substances are made in different combinations, depended on their shape. Borro assumed that the elements were not substances as such but something in between substance and accident. Borro also performed experiments with wooden and leaden objects by dropping them from a window.

In Buonamici's work "De motu" the author adhered strictly to the Aristotelian concept of motion or change in general, and he therefore disagreed with Borro. In his introduction, Buonamici mentioned that he wrote his book as the result of a wider academic discussion on the subject at his university. So when Galileo came on the scene, the debate in Pisa was already in full swing, and the Pisan professors played a fundamental role in triggering his research. In his own work "De motu antiquiora," he cited Borro's publication. Thus "De motu antiquiora" presented just one specific view in the dispute that was going on at that time already.

In "De Motu" Galileo developed a theory of dynamics derived from Archimedes' investigation of floating bodies. From the beginning, Galileo intended an alternative approach to the concepts of Aristotle. One important aspect dealt with the problem of upward motion. Another point of criticism concerned the movement of projectiles. Galileo based his explanation on the Archimedean findings on buoyancy and the impetus theory, dating from the 14th century. The impetus theory was based on the assumption that a body—once set in motion by the action of another body—gets a force impregnated on it, which continues to keep the body in motion until it arrives at an obstacle. This force, however, does not correspond to what physicists call force in our day, but was thought to have some sort of immaterial character at that time. The impetus theory was first proposed in the 6th century by Philoponos and developed further by Avicenna and later scholastics. Newton still used the word "impetus," but as a synonym for inertia.

Galileo explained upward motion through a mechanism that temporarily removes the heaviness of a body in a decreasing way. For falling bodies, he assumed the opposite: the natural lightness of a body is cancelled out by an impregnated force in an increasing way, leading to acceleration.

In his treatise, he also referred to the opinions of his Co-Professors, mentioning Borro even by name. Borro based his explanation of free falling bodies on the Aristotelian theory that their movement depends on their relative composition of the elements (air, fire, water, earth) with respect to the medium in which they are moving, e.g. in air. Averroes had developed a theory that for example air is heavier in its own environment etc. Thus, bodies containing more air than others (for example wood as opposed to lead) would fall more swiftly. Borro took up this theory to try to explain his own observations. Galileo refuted this explanation by comparing the relative weight of the four elements. On this basis, he concluded that lead should fall faster, since it was composed of more earth and water. From this argumentation, it can be seen that all scientists still adhered at that time to the ancient concept of the four basic elements in nature. Borro even claimed that there was more air in lead than in iron and therefore iron should be heavier in air. Both Galileo and his colleague Buonamici disagreed with the assumption that there was more air in lead than in iron. While Buonamici argued only on the basis of the relative content of air in lead or iron, Galileo based his conclusions on the simple comparison of the relative weights of the three substances with no regard to the speculative composition of these substances.

Both scholars were of the opinion that lighter bodies would initially fall more swiftly that heavier ones, but in the course of their descent the heavier ones would increase their acceleration. The phenomenon of acceleration was still explained by the impetus theory. Galileo himself did not advance a better explanation, although he rejected the ones of his colleagues. He declared the necessity of a completely new approach to describe the observed phenomena.

The main question remaining was that of acceleration. Aristotle claimed that the nearer a body came to its natural hierarchical position the faster it moved—accelerated. This was called "natural motion" to be distinct from "forced motion". It was assumed that objects subjected to forced motion decelerated. In his work, Galileo called this explanation by natural motion into question with a thought experiment comparing the speeds of a falling body from different heights (according to theory an object falling from a very high tower would have lesser speed somewhere along its path than the same object falling from a very low platform).

Finally, Galileo concluded that there is no such thing as "natural upward motion" as an inherent quality of bodies as Aristotle had claimed. All physical objects had a weight attached to them, and when upward motions were observed, they were due to the Archimedean Principle. As to the question whether Galileo did indeed carry out real experiments—even if not from the leaning tower—in these early years, there seems to be little doubt of that, given that his contemporaries Borro and Buonamici reported their own experiments in their respective papers.

Chapter 5
Padua—Important Steps in Astronomy (1592–1610)

Cosmologies

Ever since man started wanting to understand his habitat, he has reflected about the reality of his environment. His prime ambition was to describe this environment as closely to reality as possible. This task has still not been completed, even today. We remain captured by images or—in closer approximation—by models. Here is a short outline of the model history [6].

One of the very early cosmologies was handed down from India. It states that 4320000000 person-years correspond to one single day of the Brahma. On this one day, the whole cosmos passes through its complete cycle—again and again: every single atom dissolves in the primordial water of eternity from which everything had once been created.

On the basis of known records, the pre-Socratic philosophers were the first to develop cosmological theories outside a mythological context. And the one philosopher who presented a concise model for the creation of the world was Anaximander, who lived from around 490 to around 550 B.C. Although no original document from his own hand has been preserved, philosophers and historians who came after him handed down his findings for subsequent generations. We will discuss his theories below.

Plato, appearing after Anaximander, wrote that the world has been created in such a way that human reason is capable to understand it. This world remains forever in its original cosmic state. It is a living organism that possesses a soul and reason. The sun, the moon and some stars came into being just to enable the measurement of time.

Aristotle also declared that there had never been any proof that the world had undergone any cosmic change. He assumed that the Earth was at the centre of the world and a sphere enclosed in other spheres containing the rest of the world. During his lifetime, the circumference of the Earth was calculated with a precision of 85% of today's value. This value still served as the basis for Columbus' calculations in preparation for his voyages of discovery.

© Springer International Publishing AG, part of Springer Nature 2018
W. W. Osterhage, *Galileo Galilei*, Springer Biographies,
https://doi.org/10.1007/978-3-319-91779-5_5

The Muslim philosopher Avicenna, living from 980 to 1037 A.D. stated that

- time is a measure for movement and
- space exists only in the imagination of human consciousness and must be regarded as an entity separate from matter.

Near the end of the Middle Ages, Nikolaus Cusanus (1401–1464) concluded that all parts of heaven including the Earth were in constant movement. Shortly afterwards, Copernicus, Kepler and Galileo decided the discussion in favour of their heliocentric world model (to be discussed in more detail in Chaps. 6 and 7). At around the same time Giordano Bruno theorised that the universe must be full of uncountable suns and uncountable Earths.

Now follows a succession of researchers and philosophers—among them Huygens, Halley, Wright and Kant—who occupied themselves with the number of fixed stars, the interpretation of the Milky Way and the orientation and substance of galaxies. And in 1835, Auguste Compte was convinced that it did not make any sense at all to speculate about the composition of fixed stars since nobody would ever be in a position to verify it.

The basis for modern cosmological models are the following astronomical observations:

- The universe is homogeneous and isotropic across distances between 10^8 light-years and farther.
- Stars, galaxies and galaxy clusters move within distances of orders of magnitude of between 10^6 and approximately 3×10^7 light-years.
- From a helicopter's perspective, one could scarcely detect differences in the distribution of matter within a random volume with a side length of 10^8 light-years.
- The universe is expanding.

These facts are the pre-conditions for any modern cosmological model. To develop them, a number of illustrious scientists have played significant roles. One of them was Edwin Hubble, after whom the famous space telescope is named.

Telescopes

The American astronomer Hubble observed a spectral shift caused by the Doppler Effect due to the movements of galaxies with regard to their direction and speed. Their movements show exclusively red shift. This means that all galaxies are moving away from the Earth. It also means that the universe is expanding. Furthermore, Hubble formulated a proportional relation between the red shift and the distances of galaxies. Galaxies move away from the Earth more rapidly the greater their distance from it is. The proportional constant on which this expansion is based is called the Hubble Constant. Its value is approximately 71 [km/s/megaparsec]. Based on this relation, the age of the universe is calculated to be between 13 and 14.5 billion years.

After the final Apollo mission, the Space Shuttle became the workhorse of NASA. The three shuttles performed a total of 135 missions. One of the more successful missions released the Hubble Space Telescope into orbit. This telescope has been functioning since 1990. It has been the largest optical telescope ever launched and operates at a resolution of about 500 km per pixel, when looking at Titan, one of the Saturn moons at a distance of 1.45 billion kilometres from the Earth for example.

Hubble was one element in an array of telescopes comprising the Spitzer Telescope, the Chandra X-Ray Observatory and the Compton Gamma Ray Observatory. Spitzer was named in honour of the astrophysicist Lyam Spitzer. It was launched in 2003, and its mission was to observe events in space that generate infrared radiation. For the functioning of its detectors a special coolant had to be chilled down to −271 °C. This coolant was exhausted in 2009, and thus its main mission terminated thereafter.

In March 2009, another space telescope named after Kepler was brought into orbit to search for exoplanets, i.e. planets orbiting other stars. Because of mechanical problems, its main mission terminated, and its operation continued on a reduced scale from 2014 onwards.

In May in the same year, the Planck Surveyor Telescope was launched by the ESA. In an elliptical orbit between 270 and 1,197,080 km, it reached the Lagrange point of the Earth-sun-system. Its mission was to record the cosmic background radiation during its 1554 days of operation. In 2013, its mission terminated and the telescope was shifted into an orbit in which it cannot be captured and destroyed by the Earth's gravitational field for at least the next 300 years.

Since 1968, different nations have put more than 45 telescopes or observatories into space with different dedicated missions. Four more are planned by 2020. Taken together, the observations that they can make cover almost the entire spectrum of electromagnetic waves, of which light, visible to the human eye, is but a small section. But visible light had been the trigger and the first source of cosmic observation from the beginning.

Looking at Stars

Coming back to Anaximander, of interest to us in this context is his interpretation of stellar light. Anaximander postulated one single origin of all known objects—one singular first event (such a "singularity" is at the centre of the Big Bang Theory of our time; it also has a mathematical meaning). In his theory, this singular primordial event could not be reduced to something even more original. He called this beginning Apeiron, the Unlimited. This inexhaustible source is in contrast to the "limited" things, i.e. objects of the everyday world of change over time: seas and rivers, heat and cold, earth and stars. In contrast to these, the Apeiron never ages and will never disappear. Both exist together: the Unlimited and the things of everyday life, whereby the Apeiron does not stand for something abstract, but for something essential. Both depend on each other, whereby the Apeiron had not been created and will exist infinitely.

Furthermore Anaximander assumed the existence of elementary forces that are responsible for climatic phenomena, the seasons etc. These phenomena originate through some interaction of these elementary forces: hot dry fire and cold moist water. His approach was to deduce the emergence of these elementary forces and their relations to each other from the Apeiron itself. For this, he resorted to the idea of some kind of spontaneous creation from something like some sort of semen. In this semen, all antagonisms of warm and cold, moist and dry were initially unified. Fire surrounded everything; in the interior a dry nucleus was formed, which was surrounded in turn by a nebulous layer. In this way, an enormous pressure was generated, which made the fire crust burst. Thus an explosion stood at the beginning of Anaximander's world model—a Big Bang.

Thereafter, the remaining strips of fire were covered by fogs containing apertures, through which we can observe the fire—our stars, the stars Galileo studied some 2000 years later. In the middle of this cosmos, there remained the hardened nucleus—our Earth. Anaximander thought of it as a cylinder with us living on the top of it (more of Anaximander's theory in Chap. 7).

Galileo in Padua

In 1592, Galileo obtained a chair in mathematics at the university of Padua, initially only for 6 years. His assignment was prolonged in 1599 for another 4 years with the option of a further two-year extension thereafter. His salary amounted to 160 Ducats per annum. He started lecturing in December on the subjects of geometry and astronomy. His contract was again extended in 1603 for another six years at 320 Ducats.

The other person interested in this position at that time had been Giordano Bruno. Giordano Bruno was born in 1548 and burned at the stake in 1600. He belonged to the Dominicans and had a keen interest in astronomy. This led him to develop his own theories about the cosmos and the stars. In his thinking, he went far beyond the model of Copernicus. For him, the fires that Anaximander saw through the primeval fog were nothing less than other stars like our own sun. In consequence, the centre of the universe was not our sun, but there was no centre at all, and the universe was infinite. It is a matter of debate whether this altogether very modern concept was the main cause that eventually lead to his trial and execution, since Bruno had at the same time also formulated doubts about other central dogmas of the Catholic Church. Nevertheless, his fate illustrates the difficulties that other scholars and scientists also faced during Galileo's lifetime.

Galileo remained in Padua for 18 years. While he was teaching there, the town was ruled by the Venetians and had been so ever since 1405 (and continued to be so until 1797). His financial situation had initially improved significantly in comparison to Pisa. Nevertheless, he gave private lessons to some students and engaged in some businesses by selling his compaso—a predecessor of the slide rule. For the construction of this and other instruments, he later employed his own mechanic, Marc Antonio Mazzoleni.

Inventions

His biographer Viviani claimed that Galileo also invented the thermometer. This was not quite the case. Galileo discovered the principle according to which the thermometer later named after him functions. A tube is filled with some liquid containing small floating bodies made of glass (balls) in it. If the temperature of the liquid rises, it expands and thus its density diminishes. The balls, having a relatively higher weight, begin to sink. At lower temperatures, the liquid contracts and increases its density, so that the balls rise up. The Galilean Thermometer works within a temperature range between 18 and 24 °C. Another invention dating from that time was the proportional divider or sector, also called the mathematical compass, about the invention of which a dispute on priority with Baldasare Capra erupted, won by Galileo before a tribunal in the year 1607. Such a sector is used to divide geometrical distances into certain proportions and to enlarge or diminish them. At that time, its main use was in the military field.

In 1609, Galileo constructed his Occhiolino, a composite microscope with a convex and a concave lens. Although the Academia Nazionale dei Lincei in Rom claimed that their countryman was the first inventor of this instrument, it had already been presented by the Dutchman Zacharias Janssen at the Frankfurt trade fair one year earlier.

Galileo's financial situation grew worse again after the death of his father. His father had in the past supplied him occasionally with clothes and even books. Now he himself had to take care of his mother and several of his siblings. He had also started a relationship with Marina Gamba from Venice, with whom he had three children: Virginia, Livia and Vincenzo. The two girls were later placed in a monastery, since—being illegitimate—they never would have had a chance for a decent marriage. His son joined him later in Florence after Marina Gamba married Giovanni Bartoluzzi.

Scientific Instruments

But Padua was the place in which Galileo really started once and for all to revolutionize mankind's view of the universe. His interest in astronomy was triggered by the appearance of a "new star" in 1604—in fact a supernova. In 1605, a small paper was published under the title "Dialog about the New Star" under the pseudonym Ceccio di Ronchitti. It is widely accepted that the true authors were Galileo and his pupil Girolamo Spinelli. The dialog between a scholar and a naïve counterpart dealt with the difficulties of maintaining the scholastic view of the immutability of the heavens. It was a first timid approach to Copernican ideas and a sort of humoresque predecessor of the famous dialog about the world systems, which would later be one of the causes for Galileo's break with the traditional worldview of the Catholic establishment.

To arrive at his astronomical discoveries, to be described further on, he had to equip himself with an instrument that suited his purposes. This he accomplished by improving the existing invention of someone else. This invention belonged to a set of technical apparatus essential for the development of natural science on its path to modern times, including the microscope, the thermometer, the barometer, the air pump and the pendulum clock.

The Galilean Telescope

The instrument in question was the telescope, invented by the German spectacle maker Hans Lippershey (1570–1619), although even he may not have been the first to look through a tube "to see things far away as if they were nearby". In fact, he did not obtain a patent, for which he applied in 1608, for his telescope, because other members of his trade had filed for patents at around the same time for similar inventions.

Lippershey was born in Germany near the German-Dutch border, but lived most of his life until his death in the Netherlands and became a Dutch citizen. There exist several different stories about how he came to build a telescope, also coined "the Dutch perspective glass". According to one report, he simply copied one that had been invented earlier by someone else. In any case, his telescope used either two convex lenses or one convex objective and a concave eyepiece. The latter version produced an upright image to the viewer. Its magnification was 3X.

Soon after Lippershey's patent application, the design of his instrument was disseminated all over Europe, which caused other people to improve it. One of them was the Englishman Thomas Harriot (1560–1621). This astronomer was the first person to look at the moon through a telescope and to draw a picture of this somewhat enlarged heavenly body—even before Galileo did so. He also detected sunspots one year later in 1610. Just as the invention of movable letters in printing furthered the dissemination of Luther's reformation and thus was invented just in time the invention of the telescope became crucial for substantiating the Copernican world model, as well.

In the end, it was Galileo, who first perfected the telescope. He constructed it from lenses he could buy on the market. Initially its magnification was 4X, later 8X and even 33X. He demonstrated its use to the government of Venice on the 25th of August 1609. The members of the government were deeply impressed. Galileo's primary use of the instrument was the observation of heavenly bodies. But he had written to Leonardo Donato, the Doge of Venice, previously, praising his invention and its qualities in quite another context, to arouse the interest of a man in power, individuals

optical beam path

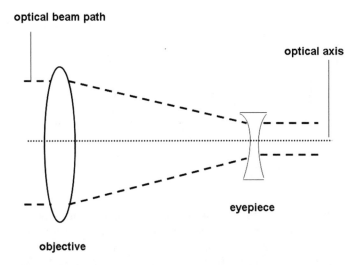

optical axis

eyepiece

objective

Fig. 5.1 The Galilean telescope

who generally have other designs than astronomy. On the 24th of August, he wrote that it could be useful to any business on land or water, for example discovering enemy ships much earlier, when they were approaching—about two hours earlier than without a telescope. This would enable the user to judge the size of the enemy fleet and its capabilities in order to take early measures to defeat it. On land, his instrument could be used to spy on enemy fortifications and movements.

Figure 5.1 illustrates the functioning of the Galileo Telescope.

The device had a collecting lens as an objective and a diffuser lens of small focal length as an eyepiece. The focal points of objective and eyepiece coincide on the side of the observer. The telescope has only a small visual field, but represents objects in an upright position and not reversed. Today, Galileo's design is still used in opera glasses. Since the ocular has a negative focal length, it has to be positioned within the focal length of the objective. There is no intermediate real image produced. The advantages of the Galilean Telescope are to be found in its short length and in its upright image. Its disadvantages are the small visual field and—in comparison to Kepler's Telescope—the difficulty of locating observed objects precisely, since the application of cross hairs is not possible because of the missing intermediate real image.

Johannes Kepler had presented the design of his telescope in his publication "Dioptice" in 1611. His instrument consisted of objective and ocular, both of which were convex. As a result, he initially got an inverted image of an observed object. However, it could be made upright again by employing a third convex lens.

Further improvements were achieved by Christiaan Huygens (1629–1695). Huygens invented a special eyepiece, later named after him. He used two plano-convex lenses with the planes facing towards the eye of the observer. These lenses were sep-

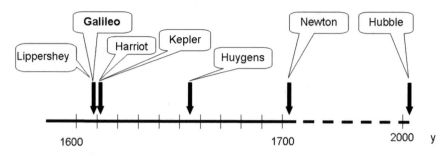

Fig. 5.2 The history of the telescope

arated by an air gap with the focal plane between the lenses. This eyepiece showed no chromatic aberrations. They were useful for telescopes with very long focal lengths.

The development of course did not stop there. Isaac Newton introduced mirror technology to telescopes for the first time so that an eyepiece could be mounted at the side of the telescope tube. More and more refined technologies lead to a steady improvement of the instrument until we arrive at today's Hubble, the most advanced successor of Lippershey's invention. Figure 5.2 illustrates the position of Galileo in this context.

The Sidereal Messenger

Galileo directed his telescope at the moon and found that the surface of the moon was rough and bumpy, consisting of heights, crevices and craters. The other planets appeared to him disc-like and not point-like as distant stars, and he thereby discovered four satellites of Jupiter and observed the phases of Venus. Although he was not the only person to have detected them, he discovered sunspots. He also found that nebular objects such as the Milky Way could be resolved into individual stars. In March 1610, Galileo published his findings under the title "Siderius Nuncius". And he started to question the age-old beliefs about the structure of the universe and finally arrived at his first break with the traditional concept of the world.

Already in 1597, he had confided in writing to a close colleague that he believed that Copernicus was right with respect to Ptolemy. He also wrote to Kepler in the same year about his views, but at the same time insisted that he did not want to make a fool of himself in public and thus would not publicly support Copernicus' theory. His letter was a response to Kepler, who had sent him a draft of his first publication "Vorläufer kosmografischer Abhandlungen, enthaltend das Weltgeheimnis über das wunderbare Verhältnis der Himmelskörper und über die angeborenen und eigentlichen Ursachen der Anzahl, der Größe und der periodischen Bewegungen der Himmelskörper, bewiesen durch die fünf regelmäßigen geometrischen Körper" (Forerunner of a cosmographical treatise about the inherited and distinctive causes of the number, size and periodic movements of heavenly bodies, proved by the five regular geometric bodies) for critical assessment.

In the same year, Galileo published guidelines about the structure of the cosmos based on Aristotelian and Ptolemaic foundations. His dualistic approach can be explained by the fact that he was not yet in possession of sufficient scientific material to definitely prove the Copernican theory at that time. At the same time, he also did not want to endanger his academic career. He maintained this ambivalent attitude for some time to come in his further professional life. Indeed, he adopted arguments from the Aristotelian school of thought to ridicule the claim that the Earth was moving, including arguments about the disastrous effect such movements would have on falling bodies and the flight of birds. Until 1610, his other writings dealt mainly with practical subjects such as the construction of military fortifications and ballistics.

All this changed when he made his own discoveries with the telescope: Jupiter was circled by its own set of heavenly bodies—its satellites. Galileo knew his discovery was outrageous. By publishing this in the pamphlet "Siderius Nuncius"—Sidereal Messenger—he broke with the two thousand year old tradition that the Earth was the only celestial body to be circled by other ones—the planets.

The Siderius Nuncius is a tract in which Galileo consolidated his astronomical findings and the conclusions to which he was in consequence led. It was primarily addressed to other scholars and not to the general public. He classified his findings as "news never before heard of", brought about by the employment of his new instrument, the telescope. As an example, he mentioned that the number of fixed stars he had seen through it exceeded the number visible with the naked eye by a factor of ten. If the telescope is pointed at the moon, then the satellite's diameter was enlarged by a factor of thirty, its surface by a factor of nine hundred and its volume twenty seven thousand times.

One of the problems he encountered was how to measure the distances between the different stars. He finally found a method by applying the intercept theorem in combination with the theorem of Pythagoras.

The Moon

He included many drawings of his observations in the text, the first being of the moon with its valleys and mountains and the Earthshine, which he was the first person to have observed and documented. At one point in his descriptions, he inserted the revolutionary remark that the Earth itself would exhibit similar features as it moved, contradicting the canonical wisdom that fixed the Earth motionless at the centre of the universe.

His explanation as to why the moon's surface features could not be observed by the naked eye and the moon's edge always looked rather smooth, however, would not hold over time, since he traced it back to qualities of an all pervading cosmic ether, the existence of which was for a long time taken to be real, until the failure of the Michelson-Morley Experiment, which in turn was explained by Einstein through the introduction of his Special Theory of Relativity.

One of Galileo's other observations was that the depths and heights of the moon's surface features exceeded those observable on the Earth significantly. With trigono-metrical calculations, he established the height of one particular mountain to measure four Italian miles, whereas to his knowledge there was no mountain on Earth that exceeded one mile in height. The drawings of the moon showed different intervals in time: at different hours during the night and during different phases as well.

Fixed Stars and Milky Way

One of the riddles to Galileo was the fact that the fixed stars could not be enlarged by his instrument in the same way as the moon. He assumed that the radiation of the stars was caused by some kind of hairy rays, if the stars themselves were indeed some sort of small balls. In subsequent drawings, he supplemented existing zodiacal signs, adding his newly observed fixed stars to them.

A further revolutionary finding was that the Milky Way was nothing but an accu-mulation of an uncountable number of fixed stars, as were some other cosmic objects known until then as nebulae.

Jupiter's Satellites

Galileo's most important discovery with respect to the existing world model was the four satellites encircling Jupiter. In a set of sixty-five meticulously fabricated drawings, he detailed the positions of these satellites from the 27th of January to the 2nd of March in 1610. He could not but take this as further evidence for the Copernican theory, which placed the sun at the centre of the universe. Jupiter was encircling the sun and was at the same time encircled by four "planets" just as the Earth was encircled by its moon, and both Jupiter and Earth were moving around the sun. Shortly afterwards, the Jovian satellites were seen by Simon Mayr from Gunzenhausen in Bavaria, at that time court astronomer in Ansbach. Galileo accused him of being a plagiarist, but Marius was later rehabilitated.

Galileo had dedicated the "Siderius Nuncius" to Cosimo II Medici, Grand Duke of Tuscany, and named the four satellites of Jupiter the "Medicinian Celestial Bodies". This dedication, however, had been initiated by Belisario Vinta, first secretary of the Grand Duchy of Venice, who was an acquaintance of Galileo, and to whom he had communicated that he was in Venice to supervise the printing of the Siderius. Galileo's original idea had been to name the Jovian satellites "Cosmic Planets", but Vinta suggested otherwise, to which Galileo agreed. In fact, people felt honoured by the "Medicinian Stars" and in Florence a host of followers arose, calling themselves

"Galileists". They all were confident at that time that they could stand up against traditional science.

Already in the preface, Galileo stated that Jupiter was moving around the centre of the universe, which was the sun. Before publication, the text was checked by the representative of the Holy Inquisition, who found that it did not contain anything that was directed against the "Holy Catholic Faith".

It is an example of Galileo's ambiguous posture that until the publication of the Siderius, he regarded his scientific convictions as a private matter, even though he had been teaching as a professor of mathematics at Padua University since 1592.

Other publications during his stay in Padua included a "Treatise on the Sphere" for his students, a paper "The star that appeared in 1604" and another treatise on hydrostatics at the end of 1606. He had also continued his mechanical research and invented or constructed a number of other devices: a pendulum to further investigate the phenomenon of acceleration and a water pump.

Chapter 6
Florence—Discoveries and Conflicts (1610–1623)

Venus

It was on October 18th 1967 that a man-made object touched down on the surface of the planet Venus for the first time ever. It was probably also the first time ever that such an object launched from Earth reached any other planet at all. The vehicle was called Venera 4. It was part of the Soviet Venera program to explore our sister planet Venus. The space vehicle consisted of a hub and a landing module. The lander was equipped with a variety of instruments to transmit data about the chemical composition of that planet's atmosphere, its temperature and pressure. The duration of the mission lasted 127 days from launch to descent. It aborted after reaching an altitude of about 26 km from the Venutian surface at a pressure of 22 atmospheres and a temperature of 262 °C.

Generally speaking, Venus is rather similar to our Earth. It belongs to the class of planets classified as "Earth-like", i.e. consisting mainly of rock and having a thin atmosphere, contrary to "Jupiter-like" planets consisting mainly of gas. Venus possesses nearly the same diameter as the Earth, about 12,000 km, and has the same material density. The main differences, however, are to be found in the atmospheric composition and the nature of the surfaces.

Venus is the only "Earth-like" planet with an atmosphere made opaque by dense clouds. The atmosphere's main component is carbon dioxide (95%). The other constituents are: sulphur dioxide, argon, some water vapour, carbon monoxide, helium and neon. Its clouds at a height of about 75 km consist of sulphuric acid. It is assumed that the sulphur had been generated by volcanic activity during early stages of the planet's development. Clouds in the upper layer of the atmosphere move at a speed of about 100 m/s in the direction of Venus´ rotation. They revolve around the planet in four days. The total mass of the atmosphere is 90 times that of the Earth's, leading to a pressure of 92 bar at ground level.

Venera 4 (Fig. 6.1) was by no means the only space probe that tried to have a look at Venus. There have been a total of 41 attempts, starting with Sputnik 7 by the Soviet Union (1961), up until today (2017) with Akatsuki by Japan (2010); four

© Springer International Publishing AG, part of Springer Nature 2018
W. W. Osterhage, *Galileo Galilei*, Springer Biographies,
https://doi.org/10.1007/978-3-319-91779-5_6

Fig. 6.1 Venera 2 space vehicle (© NASA)

more missions are planned. Of those 41 missions 15 resulted in failures, six were partially successful (target reached, but data transmission lasting only a few minutes, for example), 19 with full mission achievement. Other countries or organisations participating in the quest for Venus were the USA and ESA (European Space Agency).

Renaissance Florence

In the year 1610 Galileo was appointed first mathematician and philosopher of the Grand Duke of Tuscany in Florence (Fig. 6.2). Florence had been ruled on and off by the Medici family. Their reign had been restored again by the Pope in 1537 after Florence had been a republic for a short period of only ten years prior to that. The Medici became hereditary dukes, and ruled the city for two more centuries.

When Galileo relocated to Florence, he cut his ties to his former housekeeper, Marina Gamba, with whom he had had three children. His new position did not obligate him to engage in any teaching. Thus he had ample free time to follow up on his astronomical researches. At the same time, he took over leading positions in the

Fig. 6.2 Renaissance florence

Accademia della Crusca, of which he had already become a member in 1605 while still in Pisa.

Right at the beginning of his residency in Florence, Galileo pointed his telescope at the planet Venus and discovered the phases of that planet. Just like the phases of the moon, the phases of Venus represent different shapes of light caused by changes in the sun's illumination, as the planet moves around it. What Galileo saw was a small 60" wide sickle, when Venus was positioned between the Earth and the sun. Similar appearances could be observed when the planet stood at right angles to the sun with respect to the Earth or beyond the sun.

Galileo communicated his findings to the Jesuit Christoph Clavius, who had made similar discoveries. The consequences of these observations were devastating for the accepted world model at that time: if Venus was exhibiting different phases of illumination, then it had to move around the sun and not around the Earth!

His scientific discoveries had reached Rome, and Galileo visited this city in 1611. He brought along his telescope and demonstrated the landscape of our moon, the four Saturn satellites, the Venus phases and sunspots to Cardinal Bandini and Jesuit scholars in the Quirinal gardens. However, this demonstration was not as straight-forward as it may seem today. Early on, adversaries of Galileo, especially one of his colleagues in Bologna, Professor Giovanni Antonio Magini, had pushed him to demonstrate the instrument to other professors, the majority of whom refused to look through it at all, while others were unable to see anything due to their inexperience in observation. This fiasco was then widely reported by Magini. But after his demonstration in Rome, the mathematicians Christoph Calvius, Cristopher Grienberger, Odo

Malcotio and Giovanni Paolo Lembo, who had been charged with examining the new discoveries, wrote a favourable expertise to Cardinal Roberto Bellarmi. This finally led to Galileo's nomination to be the sixth member of the Accademia dei Lincei. He thus became Galileo Galilei Linceo: Galileo the Lynx. During his stay in Rome, he was received in audience by Pope Paul V. His membership in the Accademia and his connections to the Florentine court and to the higher clergy, as well as his friendship with representatives of the aristocracy encouraged Galileo in his leaning towards the Copernican system.

Sun Spots

Between the end of 1610 and the middle of 1611, Galileo observed a strange phenomenon while directing his telescope towards the sun. Dark spots seemed to appear and then disappear after some time on the sun's surface. The sun was regarded as a perfect heavenly body not prone to undergoing changes or containing imperfections. When Galileo communicated his observations, other astronomers like the Jesuit Christoph Scheiner (under the pseudonym "Apelles latens post tabulam") tried to save the sun's perfectness by claiming the sunspots to be as yet unknown satellites moving around it. There was also a dispute about priority in the discovery of this phenomenon. Scheiner claimed to have been the first to have observed them, but in fact Galileo had presented them in Rome six months prior to the Scheiner's announcement.

Sunspots are caused by concentrations of the magnetic flux in the photosphere of the sun. Because of their magnetic origin, they appear as pairs and may last for a few days or even some months, moving across the surface of the sun with a speed of some hundred meters per second. Their sizes vary between a few to more than a hundred thousand kilometres.

Galileo published his findings in "Lettere solari" in Volgari, the language of the ordinary people. All of his soul searching and argumentation in connection with the sunspots was best summarized in a letter he wrote to Markus Welser on the 4th of May 1612. Welser was a German Humanist and publisher, living in Augsburg. He was a member of the Accademia dei Lincei since 1612. In this letter, Galileo frequently referred to three notes under the name "Apelles", a pseudonym obviously known to both correspondents, containing words of caution written by Welser to him.

Welser had contacted Galileo three months earlier, but due to some unknown indisposition Galileo was only then in a position to reply. Another reason for this delay was the fact that he had had to deal with a flurry of other requests on the same subject from friends and acquaintances. He emphasized that he had to be very careful before publishing any new findings because of the infinite number of enemies who were opposed to anything new at all. He had even been advised that he had better join the rest of the world in erring rather than being the only person to speak out the truth. Concerning the sunspots themselves, he confessed that he knew very well what they were not rather than what they were.

Nevertheless, sunspots were real and not a deception. He had observed them for eighteen months and shown them to a number of trusted friends in Rome. Apelles claimed that their movements appeared to be similar to those of the planets Venus or Mercury around the sun. He argued that they resided neither in the "air" of the sun nor in its body. Apelles rejected the idea that there could be any impurities in the brightly radiating sun, but Galileo would not go along with this, asserting that descriptions and attributes had to adapt to the essence of things and not vice versa. He also rejected Apelles' assumption that the alleged sunspots were in fact much darker than any dark apparition on the surface of the moon.

On several occasions in his comments, Galileo stated without any hesitation that all of the planets were moving around the sun and thus accepted the Copernican world for good—even though, in his view, there may still have been philosophers who would be deeply disturbed by this new model. He accepted the fact that nothing else could have been said in the past about the true nature of the world until the invention of the telescope.

Finally, Galileo discussed the nature of the sunspots. "Apelles" believed that they consisted more or less of the same substance as other planets or "stars", which would substantiate his claim that the spots really would be planet-like and orbit either around the sun or even around Jupiter, just like the four satellites that had been detected by Galileo. Galileo himself does not exclude the possibility that some as yet unobserved additional planets might populate the space between Venus and Mercury, but rejects the idea of identifying those with the sunspots, since in that case the spots would be permanent objects, not fading away after a certain time. He concedes, however, that the only thing about them that he knew was that they resided somewhere in the sun, but he knew nothing about their material substance. On the other hand, he excludes the possibility that they were some sort of clouds in the sun's atmosphere.

In the course of his comments on Apelles' theories, Galileo conceded on several occasions that he believes Apelles to be a man of free spirit who has already accepted the new basic structure of the universe, but who, on the other hand, still clings to traditional notions, accustomed to them by long use. On one occasion, he differentiated between those practical astronomers who had the task of calculating the positions of stars and planets for purposes of orientation and thus were bound to employ the epicycles of Ptolemy, and those astronomers, whom he calls "philosophical"—like himself—who are interested only in the true construction of the world.

At the end of his letter to Welser, Galileo excuses himself for some of his inconclusive argumentation. He accepts that the subjects they were dealing with were of such novelty and difficulty, provoking so many different opinions, leading to as many approvals as rejections, that he had become quite fearful and helpless when expounding upon his own position.

This whole letter illustrates the extent of the struggle that not only Galileo but many of his contemporaries had to go through at the turn of the time they were living in. The abandonment of tried and accustomed ideas about the world and man's position in it was a painful process. What seems to us straightforward and obvious (notions to which we are accustomed through long use) were regarded as extreme, were mind boggling and—dangerous.

On the 16th of June in the same year, Galileo wrote to his friend Paolo Gualdo in Padavo, referring to his letter to Welser. It was a reply to a previous address by Gualdo, who had informed Galileo about what was known and in circulation about the discovery of the sunspots. In the first part of his response, Galileo polemicized against his adversaries and expressed his satisfaction that the number of his followers was increasing steadily. He then referred to Welser's acknowledgement of his letter, but added that the final addressee Apelles would not be able to read it, since it was written in Volgare, a language that the latter obviously did not understand. Galileo's request to Gualdo was to have his letter translated into Latin, also for the purpose that other people "beyond the Alps" could understand his findings.

Two years later, on March 8th 1614 Galileo's correspondence took him a step further down the road to conflict with the authorities as he exchanged views with Tommaso Campanella, a controversial figure and author of "The City of the Sun", who had been convicted of heresy three times: in 1591, in 1594 and again in 1600. He had become acquainted with Galileo in 1593 in Padua, and they had communicated by letter ever since. In 1614, he wrote to Galileo from his relatively benign confinement about the importance Galileo's findings had for the entire world of science and philosophy. On the other hand, he disagreed with a number of propositions, such as the speculation that everything was made up of atoms or even Galileo's theory about the movement of bodies towards the centre of the Earth. He wrote that he himself was working on a new theology and complained that Copernicus had stolen his theory from Francesco of Ferrara, an Italian. Then he took up the issue of horoscopes. Campanella claimed that Galileo must believe in such things, since he had congratulated the Grand Duke on Jupiter's constellation at the hour of his birth. In his letter, he also put forward the very modern concept that everything in nature was connected with everything else by form and substance, a concept that only recently became part of Chaos Theory. Campanella's letter is ample evidence of the intellectual turmoil scholars were going through at that time.

In a letter to Giovanni Battista Baliani dated March 12th 1614, Galileo confirmed his conviction of the correctness of the Copernican system. Galileo's contact with the mathematician and physicist Baliani resulted from a recommendation by Filippo Salviati in 1613. In the same letter, Galileo speculated about the nature of the fixed stars and the planets. He believed that the fixed stars were radiating by themselves, whereas the planets were only reflecting the light of the sun. He also expressed again his uncertainty about the true nature of the sunspots.

Towards Conflict

All of this led slowly but steadily to Galileo's decision to opt once and for all for the Copernican world model. But in spite of his enlightened way of thinking, he continued to ignore facts that did not fit his own views, for example the discovery by Kepler that the planets travelled on elliptical rather than circular orbits. Circular movements fitted a more traditional attitude towards nature. He was also well aware of the fact that his observations were not sufficient to prove that Copernicus was right once and for all. The third competing world model at the time was that of Tycho Brahe. Brahe's model was something of a mixture of the old Ptolemaic one and the heliocentric. This model assumed that both moon and sun still moved around the Earth, whereas all of the other planets circled the sun. Thus the phases of Venus, for example, could be explained by Brahe's model.

Nevertheless, all of this thought and discourse finally and irrevocably led to a first conflict with his theological masters, although he tried to explain to his adversaries at the Florentine court that Copernicus could well be brought into line with contemporary biblical exegesis. One point of contention, for example, was the biblical report about the battle of Gibeon in the book of Joshua. By the way, the discussion of the meaning of this narrative has by no means subsided to this day.

The story presented in the book of Joshua in Chapter 10 is about a battle between the Israelites and the Amorites during which God made the sun and the moon to remain motionless to give more time to his people to win the battle. If the sun, however, was the centre of the universe and was motionless anyway, the Copernican theory of the cosmos would be incompatible with the biblical report. However, this would remain a problematic conclusion, since God, being capable of working wonders of any kind, could have employed any means to achieve that effect—even in a Copernican system.

There have been many exegetical attempts to interpret this event one way or the other. Recently Israeli scientists from Ben Gurion University in Beersheba explained it by a solar eclipse on October 30th 1207 BC. They also used a different translation for the old Hebrew word for "stand still", which can also be read as "becoming dark". Older explanations from critical scholarship identify the sun and the moon as ancient deities, and their passivity as a sign that the God of Israel forced them not to intervene in the battle.

In any case, Bernard Shaw, in the preface to "Saint Joan", remarked at one instance that "the chronicle of Joshua's campaigns (was) not a treatise on physics". And that was also the point Galileo made earlier. He tried to avoid the conflict by not playing biblical truths off against scientific findings. According to his view, scientists should confine themselves to the exploration of nature and not meddle with divine truths.

Galileo laid out this apparently contradictory attitude in broad terms in a letter to Benedetto Castelli dated 21st December 1613. Castelli was a Benedictine monk and professor of mathematics in Pisa. He was a confidant of Galileo and later helped him to formulate one of his defences. Two years later, Galileo re-iterated what he had communicated to Castelli in a letter to the mother of the Archduke, Christina di Lorena, in a further attempt to defend himself.

In his letter to Castelli, Galileo referred to a discussion they had had together with the Archduke and Archduchess and other members of the court about the famous Joshua passage. Before he elaborated on the discussion, he emphasized his principles in dealing with nature and the Scriptures at the same time. Neither nature nor the Scriptures can lie, so these truths may not contradict each other. But then he asserted that the Scriptures may not always be taken literally, since there are contradictory statements in the Bible itself in several places. Everything depends on the commentators, who must explain the truth behind biblical statements to the common people. The Holy Spirit had caused the Bible to be written in a way that ordinary people could understand, for the sake of their salvation. For this reason, some passages had an apparent meaning only if taken literally, whereas observations of nature with our God-given senses should never be doubted. In his opinion, it should not be allowable to use passages from the Scriptures to try to disprove scientific results. The Bible's only purpose was to convince people of its teachings that are indispensable for their salvation. And by the way, there is so little astronomy in the Bible that it should never be mistaken for an astronomy book, since it did not even mention the planets at all. Galileo claimed that his adversaries were always resorting to the fiercest weapons against him because they were uncertain in their argumentation themselves.

At the end of the letter, he explains why the Joshua passage could not be brought into accord with the Ptolemaic system. If the sun were moving around the Earth together with the rest of the heavenly bodies every 24 h, then the entire heavens would have had to be stopped. To him, the Copernican system gave a much better explanation of this biblical event.

Galileo's remarkable approach was way out of his time, long before the dawn of the Enlightenment. During his lifetime, theology was still the mother of all science, just as philosophy would be in later times, until finally each generation would adopt a "leading" science for its time, such as chemistry, physics, genetics or more recently neuroscience in modern times. So there was no way to avoid dealing with theology then. In fact, Galileo's letter to Castelli was later used in the proceedings against him.

Furthermore, in a letter from February 1615 to Cardinal Santa Cecilia, Father Niccolo Lorini of the Dominican Order, a teacher of church history, denounced a pamphlet from the so-called Galileists that referred among other things to this episode in the book of Joshua. He pointed out in one of his sermons that Copernican teachings were directed against the Bible. The intention of this written communication was to provide material for possible legal proceedings against Galileo, but the author begged the addressee not to use his letter officially, but rather to keep it secret.

Long before the Joshua quarrel, other defenders of Ptolemy had rallied against the Galileists, among them the philosopher Lodovico delle Colombe in a pamphlet from 1610 called "Contro il moto della terra" ("Against the Motion of the Earth"). In December 1614, another Dominican, Tommaso Caccini, held a sermon inveighing against the Galileists on the basis of Acts 1:8, and claiming that mathematics was an invention of the devil.

In the meantime, other scholars were well aware that Galileo favored the Copernican system. He was intent on placing it on solid scientific foundations. But attacks and denunciations to the Inquisition, especially from the lower clergy, increased at a steady rate. Galileo complained in a letter to Federico Cesi (Fig. 6.3), who had founded the Accademia dei Lincei at the age of eighteen, about these denunciations. At the time, he still hoped that the clerical hierarchy would react to his findings in a different way than the lower clergy.

Cesi sent his response on January 12th 1615. In this letter, he confirmed that Bellarmin had told him that he regarded Copernicus as a heretic, and that there was no doubt that the Scriptures affirmed that the Earth was at the centre of the universe and did not move. Cesi also mentioned that in certain circles of the lower clergy, mathematics and mathematicians were regarded as instruments of the devil and the source of all heresy. Galileo also tried to enlist the support of Archbishop Piero Dini, his friend and a Vatican official, in a letter from May 1615. In this letter, he almost despaired that he was not able to prove by experiment the correctness of the Copernican world model, and that his opponents were intellectually incapable of grasping even the simplest scientific foundations.

Galileo meanwhile had tried to convince members of the congregation about the Copernican system and his own beliefs. He had even succeeded in obtaining the backing of Cardinal Orsino, who intervened—albeit without positive result—with the Pope. Other members complained about the vigour and uncontrolled passion with which Galileo tried to defend his positions. In the end, the official church saw no other recourse but to act against Galileo.

On February 25th 1616, Pope Paul V. instructed the Holy Office and Cardinal Millini to summon the mathematician Galileo to warn him to abstain from the falsity that the sun was at the centre of the world etc. In case of non-compliance, the accused would be incarcerated. Otherwise, there was no formal trial of the Inquisition against Galileo. The next day, the defendant appeared before Cardinal Bellarmin in the Cardinal's Palace. He was admonished in the presence of among others the Dominican Father Michelangelo Seghezzi da Lodi to abstain from defending the Copernican system. Galileo was threatened with formal proceedings in case of disobedience. So he promised to obey. In any further discussion, he would treat Copernicus' theory as purely hypothetical and nothing else. Bellarmin later confirmed in a letter to Galileo that he had not been officially condemned by the Inquisition.

In a letter to Curzio Picchena, first secretary to the Medici court, dated March 12th 1616, Galileo reported about the excellent treatment that church officials had reserved for him. He had a walk with the Pope for three quarters of an hour, during which niceties were exchanged. Galileo complained about the malevolence of his persecutors, whereas the Pope assured him about the high esteem in which Galileo was held by himself and the whole congregation, and offered him his favourable ear for future causes. From this day on, Galileo refrained from any public discussions with respect to the Copernican world model.

Other publications dealing with this world model were also subject to investigations by the Inquisition. The works in contention included a book by Paolo Foscarini, a Carmelite, in which he tried to prove that the Copernican system did not contradict the Scriptures. Bellarmin had written a rather friendly letter to Foscarini on April 12th 1615, in which he tried to convince him that he was in error by citing all of the contemporary and ancient commentaries to the Scriptures, which unanimously confirmed the Ptolemaic system and Aristotle. He conceded that Copernicus' theory could well be regarded as a hypothesis for the sake of argument but nothing else. Other treatises about the Copernican system were equally in contention. A total of six other works that originated during the last couple of years in Italy, Germany and Ireland were listed in the Index Librorum Prohibitorum, the index of forbidden books, in a decree of March 5th 1616. Copernicus' main opus, "De revolutionibus orbium coelestum", was, for the time being, spared this fate. It was only put in "sus-

Fig. 6.3 Federico Cesi

pension". It could be quoted from under the condition that it was not representing the truth but only a mathematical hypothesis and was subject to corrections inserted in the preface formerly dedicated to Pope Paul III and at the end of Chapter 10. These corrections were delivered in May 1620. Thereafter, Copernicus' writings could be referred to once more, but again under the condition that they were cited as a purely hypothetical.

Back in 1616 in June, Galileo's adherence to his promise to abstain from Copernicus was put to the test after an oral dispute between himself and Francesco Ingoli, a theologian and astronomer, member of the Congregation of the Index. Ingoli summarized the discussion in a tract, which was published, listing eighteen scientific and four theological arguments against Copernicus, and requested Galileo to reply with his own argumentation, also in writing. Galileo remained silent.

In private, he took up the matter on his visit to Rome in June 1624. He had been received by His Holiness six times for long discussions, obtained confirmation of a pension for his son, a painting and two gold coins as presents. He also met a number of Cardinals for cordial exchanges: all in all a very harmonious relationship with the Catholic Church. During this visit, Galileo also learned that the church and the Pope in particular never really had condemned the followers of Copernicus as heretics, but had only reprimanded them for their presumptuousness. The question once again came up, whether astronomy was an issue of faith or not. Galileo left it at that and did not take any position himself, but only cited the opinions of others.

At the same time, he reflected on the causes of the tides. In a letter to Elia Diodati in Paris, a Swiss lawyer and Calvinist, Galileo communicated to him that he had calculated the tidal effects of both the Ptolemaic and the Copernican systems. But once again, he left his own opinion unsaid.

Il Saggiatore

Near the end of 1618, three different comets were discovered. Of course members of the scientific communities, among them Galileo, were asked for an explanation of these new celestial bodies. Galileo, however, initially remained silent on the subject. A professor of mathematics, Orazio Grassi, a Jesuit, delivered an explanation in a public lecture at the Collegio Romano. Grassi contended that comets are real celestial bodies moving beyond the moon. He based his explanations on the world model of Tycho Brahe. In Brahe's system, all planets rotate around the sun, whereas the sun and the moon orbit the Earth. Grassi's lecture was also published and disseminated in early 1619 under the title *"On the Three Comets of the Year MDCXVIII. An Astronomical Disputation Presented Publicly in the Collegio Romano of the Society of Jesus by one of the Fathers of that same Society"*.

Galileo was pressed by prominent people and friends to comment on this paper. He did so, but not in person, instead through the services of one of his pupils, Mario Guiducci, who delivered a lecture in Florence drafted by his master in June of the same year. This lecture was published under the title *"Discourse on the Comets, by Mario Guiducci, delivered at the Florentine Academy during his Term as Consul."* In this paper, Guiducci (Galileo) refuted the scientific argumentation of Grassi, since Galileo took the comets to be optical illusions.

Grassi did not leave it at that, but countered with a tract called "Libra astronomica ac philosophica" or *"The Astronomical Balance, on which the Opinions of Galileo Galilei regarding Comets are weighed, as well as those presented in the Florentine Academy by Mario Guiducci and recently published"*, which he published under the pseudonym Lotario Sarsi Sigensano in October. In June 1620 Guiducci himself replied to this publication in a letter to Grassi.

In January 1621, Galileo was elected Consul of the Accademia Fiorentina. He now was sufficiently enraged by Grassi´s latest publication on the subject that he finally sat down to elaborate a lengthy response to Grassi and lay down his principles about scientific methods. He finished the manuscript of Il Saggiatore (The Assayer) in October 1622 and sent it for review to the Accademia dei Lincei. In February 1623, his paper was cleared by the censors (Fig. 6.4). It was published in October of the

Fig. 6.4   Franceso Villamena's frontispiece for "The Assayer" Rome 1623

same year. Thus about four years elapsed from the time of the discovery of the comets until Galileo's statement. Galileo dedicated the script to Pope Urban VIII. This Pope, the former Cardinal Maffeo Barbarini, had been a patron of the Accademia and close friend of Galileo. During a visit to Rome, Galileo was assured of the continuing support of his old friend, now at the highest level in the church.

Il Saggiatore is not so much a scientific publication in that it did not contain anything new concerning discoveries or theories. From the point of view of science, its most important contribution is Galileo's firm stance regarding scientific methods. On several occasions in the paper, he confirmed that nature could be understood only if rational deductions and experimental verification are employed to investigate it. He rejected the scholastic approach of deducing scientific facts from religious truths or relying purely on ancient philosophical models accepted as unquestionable foundations up until then.

On the subject of comets, however, he was wrong, since he took them to be some kind of illusion. And there was of course nothing new on the Copernican question, which Galileo left out for reasons cited above.

On the other hand, the work as a whole shows up Galileo's character. It oozes with sarcasm, vanity and pettiness. This may have been the standard of the time, but questions of priorities for discoveries seemed to have dominated a fair part of the discussions going on all the time—much the same as in our own era.

These are the salient points in Il Saggiatore:

Right at the outset Galileo complains about the jealous and the grudging, who steal his achievements or deprecate his publications. He continues in this vein, citing the reception of his enhancements of the Archimedean principle as an example, furthermore the attack on his letters on sunspots.

He then turns to plagiarists and singles out Simon Mayr of Gunzenhausen in Germany, who claimed to have seen the four moons of Jupiter at about the same time as he himself. In an elaborate argument he ascribes the coincidence of dates between his own and Mayr's observations to the use of two different calendars (he supposes that Mayr—as a Protestant heretic—would not have used the Gregorian calendar). Mayr was in fact later rehabilitated, and a high school in Gunzenhausen is today named after him.

His closest friends had advised Galileo to abstain from any quarrel with his adversaries, which he did for a while, but in the end could not help himself from once again responding in writing. He then turns to Grassi.

Although it is obvious to him that Grassi and Sarsi are one and the same person, he denigrates the "name, unheard of in the world" of Sarsi, and generally strikes out at people masking their identity for whatever objective, and thus sees no reason to hold back his temper or to weigh every word in his reply.

He then recounts his version of the exchange on the comet mentioned above. Galileo claims that Guiducci had been present at his sickbed at the time the comets were observed and that he, Galileo, had offered some verbal explanation of this event, which Guiducci had then included in his lecture without mentioning this private communication. Galileo even went so far as to play Grassi off against Sarsi's publication, although both names refer to the same person. This continues throughout the tract. The game is: Guiducci is a respectable person, but Sarsi is but a mere imposter. In fact, Galileo remarks at one instance that Sarsi should leave him and Guiducci alone.

During this tirade, Galileo then inserts a comment, which testifies to his entire scientific approach and is really the basis of his life's work:

> Philosophy is written in this grand book—I mean the universe—which stands continually open to our gaze, but it cannot be understood unless one first learns to comprehend the language and interpret the characters in which it is written. It is written in the language of mathematics, and its characters are triangles, circles, and other geometrical figures, without which it is humanly impossible to understand a single word of it; without these, one is wandering around in a dark labyrinth.

Further, on, Galileo rejects Tycho Brahe's explanation of the comets. Brahe had been quoted by Sarsi. But even more—Galileo does not stop there, but derides Brahe as a man who could not extricate himself from his own state of mind under any circumstance.

Then Galileo turns to the subject of the discussion itself: the comet. His basic assumption is that a comet is just some sort of appearance, although he admits that he has yet no explanation how this comes about. His adversaries think that a comet is a real heavenly body, and Sarsi bases his belief among other things on ancient writers. The latter fact leads Galileo to again deride those scholars who deduce their theories from ancient scriptures. He counts himself among the few enlightened men with few followers, up against the mass of the unenlightened and their masses of followers.

He tears apart Sarsi's paper bit by bit by trying to prove logical inconsistencies within in and by splitting hairs in picking out phrases that seem vague and imprecise, like "irregular path" or "infinite number of stars". For Galileo such wording only confirms that Sarsi does not know what he is talking about.

The next round is about who first invented the telescope. Whereas Sarsi contends that Galileo had adapted the instrument from some other source, Galileo insists that he invented it himself. (In the previous chapter, the true story is reported). He even turned the sequence of events upside down: he did not discover and investigate astronomical objects because he had the telescope at his disposal; he claims that because he wanted to make astronomical discoveries he invented the instrument in the first place. And he had not been inspired by the work of any other man (the spectacle maker for example), but had deduced the workings of his invention by pure logical reasoning about the nature of optical lenses. He then expounds upon "enlargement", the length of a telescope tube, and comes back to presume that a comet is something like a toy planet, comparable to images of the sun or the moon seen in a pond of water.

More seriously, Galileo admits that there are phenomena in nature that can possibly not be explained, since human senses are limited, and that therefore he himself also has no explanation for the substance of comets. This is especially the case concerning the comet's tail. Sarsi had claimed that Kepler had refuted Galileo's stance that comets were optical illusions, which Galileo denies.

Galileo then raises some aspects concerning general propositions about the roughness of the surfaces of heavenly bodies and the movements of the planets with regard to what he calls "third motion".

The final section is once again of major interest concerning Galileo's closeness to modern physics. It deals with heat and motion and especially with the nature of heat itself. At the centre of the argument stands Aristotle's position that motion is the cause of heat, defended by Sarsi. Basically, Galileo agrees with this, but argues that Aristotle drew the wrong conclusions—for example that an arrow shot from a bow would be set on fire by its motion through the air. But Galileo himself drew the wrong conclusions by any modern standard in his argumentation as well, although he came pretty close to the mechanical equivalent of heat. The argument revolves around a presumed loss of weight in a body that is radiating heat. Galileo supposes that heat is transported by some sort of fire particles and thus agrees with the idea that motion is the cause of heat. He even goes so far as to speculate that light is created by such fire particles, at the level of atoms. But he admits that he really does not understand how all this comes about.

It would take another 217 years before Julius Robert Mayer, during his voyage on the vessel "Java" to Batavia in 1840, contemplated a possible connection between motion and heat, while watching the waves of the sea rolling against the hull of the ship he was travelling on. After his return home to Heilbronn in Germany, where he made his living as a physician, his considerations resulted in what was later called the "First Law of Thermodynamics", the foundation of the conservation of energy. However, Mayer had to fight all his professional life to get his ideas recognized, in the course of which he was even forced to spend some time in an asylum (in Chap. 9 more details of Mayer's contribution will be presented).

And it took more than another 60 years until Max Planck and somewhat later Niels Bohr found the connection between the de-excitation of electrons in atoms as an explanation for the emission of light.

Il Saggiatore was the fruit of many years of creative reflection and was dubbed the "Manifest of the New Sciences". It definitely served as a precursor of the "Dialogo" (Chap. 8).

Grassi's paper was taken up later by Jesuit astronomers and developed further in a book by Giovan Battista Riccioli called "Almagestum Novum" ("The New Almagest", Almagest s. Chap. 7) as a theory largely based on Brahe's model.

Intermediate Assessment

What if the story had ended here? If Galileo had continued his quiet research work, exploring objects in the sky and publishing on them, contemplating further on the laws of motion and exchanging views with his contemporaries, cultivating his position in society and his relations with the clerical hierarchy? Or, if he had suddenly disappeared because of illness or accident? What would the historical assessment of him be today in this case?

First of all, there is no doubt that he was a brilliant scientist and natural philosopher. Just like Baliani or Salviati or others of his time.

Then, he improved the telescope—just as Lippershey and Harriot did.

He discovered the Venutian phases, as Simon Mayr did.

And he observed the sunspots, as Scheiner did.

He did not propose a new cosmological model. That had been done by Copernicus, but he sympathized with it. When it came to the crunch, he renounced the model and held his peace in public, preferring to stay on good terms with his clerical friends.

So, perhaps today he would be remembered by a small circle of specialists as one more brilliant and vain researcher at the dawn of modern times, just like Cesi, the founder of the Accademia dei Lincei. The Galileo we remember today had still to come into existence.

Chapter 7
Cosmological Excursions

One of the greatest mysteries and at the same time a great facilitator is our keeping of time in sun years. Imagine other cosmic constellations without the Earth's rotation or orbit around the sun: without this natural world clockwork, we would indeed ride on a linear time beam, and if there were no alteration between day and night, we would have no basis even to calculate the duration of a second. We would be anchorless without a time horizon in a continuum without hope and without beginning or end. We probably would not have developed any sense of time such as we know today.

But even if things were as they are: if we would not divide the year into months and the months into weeks (leaving aside the problems of synchronising the sun year with the lunar year and the introduction of a seven day week, which has no astronomical basis)—why is it necessary to reset the calendar to 1, once the 365 days of the year have passed? Why not just continue counting, since there are no technical difficulties to prevent this nowadays? Well: at the 16244th day after introduction of the New Common Era (NCA) someone bought a new car and scrapped it then on the 17001st. Such an accounting would spare the celebration of recurring birthdays and neutralise any age discrimination.

So: what is, what was at stake and why?

It is obvious that the movement of celestial bodies and its observation carry more than just the information about the change in their relative positions between them with respect to an observer on Earth. Around and about these constellations whole systems of thought and philosophy were generated. These interpretations were so powerful that they ended up dominating the reasoning and action of mankind up to the present: payments of rents and salaries are still coupled to the passages of the moon and so are the quarterly reports of publicly held corporations.

Until now in this book, we have presented discrete and dispersed notes on different world models that are incompatible with one another. These models were touched upon briefly in the context of Galileo's discoveries and his conclusions about which of them were in his judgment correct—basically regarding the competition between the Ptolemaic and the Copernican systems. From the conflict in his own mind, which was only a reflection of the same conflict in the outside world of science, religion and politics at large, the enormous importance of these questions became obvious—apart

© Springer International Publishing AG, part of Springer Nature 2018
W. W. Osterhage, *Galileo Galilei*, Springer Biographies,
https://doi.org/10.1007/978-3-319-91779-5_7

from all practical considerations. The very lives of people were at stake, when one uttered one's interpretation of why the sun was moving in this way and not in the other. We are at the crossroads from one epoch to another, and the shaking of old foundations reverberated in the minds of the sharpest scientists of their time.

To put it all into perspective, it is useful to position Galileo on the temporal axis that leads from the earliest known systematic model building to the (probably only temporary) standard model of our own time.

Anaximander

Anaximander belonged to the Pre-Socratics. We have already heard of him in Chap. 5. Some people believe that philosophy started with the Pre-Socratics [7]. In any case, their importance is founded upon the fact that they were the first to formulate many important scientific and philosophical questions at all. This does not mean that they were also capable of answering the questions about the ultimate and penultimate things that are relevant to us today. But their critical and rational approaches served as building blocks for future thought.

These philosophers were collectively known as the "Pre-Socratics", the thinkers who had an impact on science and philosophy long before Socrates. Their role was enhanced by their reception by Plato and Aristotle. They lived during the first half of the sixth century BC on the western coast of what is now Turkey and which was then colonised by Greeks, in so-called Ionia. Later some of them relocated to southern Italy. They lived at about the same time as the biblical prophets. Philosophy itself was introduced in Athens only in the middle of the fifth century.

Turning to philosophical contemplations led to the abandonment of mythological explanations for the world, which at that time were predominant. But at the same time, it meant resorting to such conceptual expressions as "primitive state", "uncertainty of powers" etc., since philosophical language did not yet exist. Mythology had therefore been indispensable for the emergence of a philosophical attitude as such. Concurrently, important social and technological developments took place in society. These were also necessary but not sufficient conditions for the emergence of philosophy. In any case, philosophy did not emerge as a broad movement but only in the minds of a few individuals. In spite of this, the new ways of thinking were accepted within a relatively short time.

The transition from mythology to rational explanations can be explained simply with the example of solar eclipses. A solar eclipse had been interpreted as an expression of wrath of the Gods. But ever since Thales succeeded in forecasting the eclipse of 585 BC, this interpretation became obsolete. Nevertheless, pre-Socratic theories can still be characterised as speculative, supported only partially by experience.

A new term appeared: the idea of the "natural process". It was again Thales, who conceived of it first. But a true alternative with respect to mythological explanations of the world was offered by his successor, Anaximander. Anaximander continued to develop the concept of a natural process and applied it to the whole complex of questions surrounding mythological cosmology. He tried to explain the basic process out of itself. Observed events are no longer the result of mystical powers but necessary results of previously existing conditions.

For Anaximander, the existing world order was the consequence of an inevitable Big Bang, the explosion of a nucleus that had come into existence in an inexhaustible ultimate source, having been formed from the hot and dry force fighting the cold and wet force. The rupture of the ring of fire around this nucleus became the cause of all heavenly bodies and their positions. The Earth remained immovable at the centre of the cosmos. The same forces that lead to the creation of the universe would also one day be the cause of its doom.

Anaximander's approach was completely novel for his time. For the first time, the facts of the world as he found them had become questionable. He then retreats one step and creates a speculative theory as to how the observed experiences could have been brought about from certain preconditions. To understand him, creativity and a certain rational imagination, as well as rationally interpreted judgements, are required. His scientific construct of ideas had definitely been stimulated by Thales, although it is not possible to prove a classical teacher-pupil relationship. By influencing succeeding thinkers, a tradition of critical discussion was founded, constituting the core of scientific-philosophical activities.

Aristotle declared that amazement and astonishment about seemingly obvious occurrences were the beginning of philosophy. To formulate an explanatory theory against such a backdrop is quite a challenge. To deliver explanations, one has to differentiate them conceptually from experience. The discussion that then follows finally centres on the meaning of experiences and their verification of such a theory, and, from the opposite point of view, on the interpretation of experiences by the theory itself. In this way, the safety of a stance in which the question "How?" about a first explanation for an orderly universe would never be asked, was abandoned. This question "How?" had been asked for the first time by Parmenides: how could the beginning of today's world have emanated from something completely different?

Different Pre-Socratics gave different answers, leading in turn to more scepticism in wide intellectual circles. This was particularly exemplified by the thinking about non-reducible items such as atoms and elements. There was discord about the explanatory models, and generally acceptable criteria did not exist.

Another novelty introduced by the Pre-Socratics was the fact that people started to record their thoughts on paper—and on top of this, to do so in prosaic language. All mythological writings about the world had been—if at all—written down in poetical fashion. The Pre-Socratics new practice meant that important statements claiming certain factuality—even about the highest subjects—could be formulated in prose.

How do we know about the body of thought of these people? None of their works has been completely preserved. Fragments have been passed on by Simplikios, in the 5th century AD, as well as by Sextus Empiricus in the 2nd century AD. Both of these authors passed down fragments as quotations. Besides these quotations, there are also many recitations from the literature of specialists in antiquity. As is well known, Aristotle worked also on the history of philosophy. His declarations concerning our subject are indispensable, since they are reliable, just like excerpts from "The History of Natural Philosophy" by his pupil Theophrast, at the end of the 4th century BC. The most important Pre-Socratics besides Anaximander are: Thales, Anaximedes, Pythagoras, Xenophanes, Heraklit, Parmenides, Zenon, Empedokles, Anaxagoras, Leucippus and Democritus.

Turning to Anaximander himself: he was called the first systematic. According to the chronograph Apollodor, who lived in the second half of the 2nd century BC, Anaximander was 64 years old in 547/546, the year before the fall of Croesus' residence Sardes—and thus 25 years older than Thales. According to symbolic connotations current at that time, this could be seen as a classical teacher-pupil relationship.

Contrary to Thales, who developed his ideas in different, discrete areas of interest, Anaximander realised one big comprehensive design. The basic questions touched upon were: What is the origin of the Earth? What is the origin of water and stars? How can the regularity of solar eclipses be explained?

Anaximander postulated one origin of all experienced objects—one common primal cause. This common cause should not necessitate the reduction to something even more primal. He called this beginning ("Arche" in Greek) the Apeiron, the unlimited. This inexhaustible source stands in contrast to the limited things that are subject to changes over time in everyday life: the sea and rivers, heat and cold, earth and stars. Contrary to them, the Apeiron never ages and cannot perish. Thus, there exist the unlimited source and the limited things of everyday life side by side, whereby the Apeiron is not something abstract but something essential. But both presuppose each other mutually, where the Apeiron has not been created, but is eternal as well.

In addition to these ideas, he assumed the existence of elementary forces that are responsible for climatic phenomena, the changing of the seasons and so on. These phenomena were said to be caused by the interaction of the elementary forces: hot dry fire and cold damp water. His approach was to deduce the generation of these elementary forces and their interrelations from the primal source. For this, he resorted to spontaneous creation out of something primal like semen. This semen contained all contrasts of warm and cold, damp and dry—initially united. Fire enclosed everything, in its interior a dry nucleus was formed, which was surrounded by a kind of nebulous layer. In this way enormous pressure was built up, which lead to the bursting of the fiery cortex. In this way, at the beginning of Anaximander's world model there was an explosion—a Big Bang.

Thereafter, the remaining strips of fire were enclosed in fogs, leaving apertures through which we can observe the primeval fire—the stars. In the middle of this cosmos, there remained a hardened nucleus—our Earth. Anaximander supposed that this was cylindrical. Mankind lived on the upper surface.

One of the major achievements of this Pre-Socratic was that he made the cohesion of his structures accessible through the conceptual language of mathematics. Just for this reason, his pioneering works point to the most distant future. And since all moisture would one day be used up, the cosmological process would also have to run backwards, and the elementary forces would sink back into the Apeiron: the Big Crunch.

Only one single sentence of this philosopher about the elementary struggle of the natural forces has been passed down to us. With respect to the cosmological process, we can only see one tiny, nearly static detail in time. The process executes in an orderly sequence over time. Thus, between the primal semen and the decline, there exists the magnificent structure of the cosmos with everything that populates it. Besides his mathematics, there are these further legacies of Anaximander:

- The inevitability of natural processes
- The discovery of the physical concept of time together with
- Physical causality.

The natural process is time-dependent. This process is structured by a series of causes that produce effects, each cause produced by another cause and in turn producing another cause as a consequence, each part of this causative chain occurring within a limited span of time.

Since a complete text passage of this philosopher has not been preserved, this is a selection of quotations from different sources:

Anaximander, son of Praxiades from Milet. He said, origin and element are the unlimited; he did not determine it as being air or water or something similar. And these parts convert, but the universe remains unchangeable.

(Diogenes Laertos)

Opposites are the hot, cold, dry, damp etc.

(Simplikos)

There are of course those, who determine the unlimited in such a way, i.e. as something apart and beyond the elements, wherefrom they let the elements being created, and determine them as nothing else than air or water, to prevent that if one of them should be infinite the others do not perish. Because the elements nurse the relation of enmity between them; air for example is cold, water is damp, fire is hot. If one of them would be infinite, the rest would have perished long ago. Thus they say that the unlimited is something else than elements, out of which they are created.

(Aristotle)

….the grandeur of just the unlimited, because it being the all-encompassing and includes everything.

(Aristotle)

Anaximander, son of Praxiades from Milet. He identified the principle of existing things as having a certain nature, the unlimited, and from this, the worlds and the orders included in them were created. It would be eternal and non-aging and encompass all regular worlds as well. He talks about the time, because the creation and the being and the decay have been delimited in an exact fashion. He thus allocated the unlimited both as the origin and as element of existing things and used for the first time the designation origin, principle. He added that movement is eternal and that for this very reason worlds were created by this movement.

(Hippolytos)

Anaximander from Milet says the origin or beginning of existing things is the unlimited, because from it everything is created and towards it everything decays. This is because an infinite number of worlds are produced and again decay into that from where they came. He even stated the reason, why it is unlimited: because the factual creation does not relent in any regard.

(Aetios)

Anaximander, son of Praxiades from Milet, successor and pupil of Thales, states, beginning and element of existing things be the unlimited, whereby he was the first, who has introduced the term beginning. As such he did neither denominate water nor any other of the usual elements, but some other, unlimited essence, from which all universes as well as the cosmic orders contained in them are created: 'The existing things out of which the existing things are created are the same into which their decay takes place, as it should be, because they carry out justice and punishment against each other for wrong according to the temporally order' as he expressed it in metaphorical words. It is clear that he could not approve that one of the elements should be determined as the underlying, when observing the conversion of the elements into each other, but that it was necessary to introduce something besides and a part from them. In his opinion the creational process is not determined by the conversion of an element but by the secretion from it through the movement of the eternal.

(Simplikos)

There are two aspects that make these texts unique:

- The reduction to one single initial cause for the creation of the world.
- The explanation of the existence of things through the interactions of their extremes.

At the same time, Anaximander introduces the notion of something unlimited. These are very modern concepts, which are understood immediately by people of our time. But during his time, these notions were revolutionary, since up until then the creation of the world had been deduced from something mystical or from the Gods. At the same time, the question as to whether his contemporaries understood these distinct concepts differently than we do, used, as we are, to the Big Bang, or not, cannot be answered. In spite of that, we may still read these concepts according to our modern understanding.

At least his model of the origin of all corresponds much more closely to our present cosmology than many intervening models that followed it:

1. There is a beginning.
2. There is a time flow, to which events can be related; this time flow has only one direction.
3. There is movement, which continues forever.
4. The world has no boundaries.
5. Everything observable follows an order inevitably corresponding to initial conditions; a different order is not possible.
6. The unlimited will one day collapse.

Up to this point, these statements can be brought into line with the current standard cosmological Big Bang model, the expansion of the universe and a hypothetical Big Crunch. The difficulties start with the equilibrium hypotheses—i.e. that the extremes keep their balance. In his reliance on the ancient concepts of the elements fire, water, earth and air, he remains a child of his time. A direct transcription to our state of knowledge is difficult if not impossible.

But states of equilibrium are known from thermodynamics as well as, for example, in initially separated and later combined adiabatic systems. Insofar, the assumption that everything has to do with temperature points in the right direction.

In typical Pre-Socratic fashion, Anaximander sees a connection between the elements and the movements of the eternal: everything is flowing. In part, this comes close to Chaos Theory, which makes statements about the tiniest causes. Under the condition that everything is linked with everything else, a tiny impulse can cause a major catastrophe.

For our purposes, it is not only important to find out where there are analogies—analogies in the end between two world models, of which the first does not know the second, but the second, today's, knows the first quite well. The other important aspect is: how did this construct of thought develop further, who was stimulated by it? What changed after Anaximander?

Aristarch

Aristarchus of Samos (310–250 BC) did not start out to propose a new world model, but occupied himself initially with the practical details of measuring astronomical distances. Only after contemplating these did he then draw conclusions about the structure of the cosmos at which he had been looking. He wrote his observations and reflections down in a text entitled "About the Quantities and Distances of the Sun and the Moon".

Aristarchus belonged to those Greek natural philosophers who based their explanations of nature not on divine or mythological grounds but who tried to look at the world "with the eyes of science". He believed that the sun was a great fire. And he recognized that the moon did not possess its own light but received it from the sun. He concluded that the radius of the moon corresponded to that of the Earth by about one third, that the moon was at a distance of about 20 times the radius of the Earth, that the sun was approximately seven times larger than the Earth and twenty times further away than the moon.

From these proportions, he drew some interesting conclusions: if the sun is seven times larger than the Earth then it seemed unreasonable to assume that the small Earth should be at the centre and the large sun would circle the little Earth. Aristarchus was of the opinion that the big sun was at the centre and the small Earth circled it. He thus became the first person to argue for the heliocentric world model.

Aristarchus' method for determining the ratio of the distance between the Earth and the moon as opposed to the distance between the Earth and the sun used a triangle made up of the angle between an observer, the moon and the sun at half moon (90°) and the angle between moon, observer and sun, which he measured. He derived the ratio between the size of the moon and that of the sun from the fact that at a total eclipse, the moon would just cover the sun completely. He derived the ratio between the diameter of the moon and the distance between Earth and moon and the diameter of the sun and the distance between Earth and sun from the angle they covered of the partition of a sign of the zodiac. Furthermore, he derived the ratio of the moon's diameter to that of the Earth by observing the movements of the moon's shadow across the Earth during a lunar eclipse.

From all of these relative data, he finally was able to calculate the desired quantities. Because of the measuring methods at the time of Aristarchus, his results diverged far from the values accepted today. Nevertheless, his achievements can be regarded as major milestones in the history of astronomy.

Eratosthenes

Eratosthenes of Cyrene (today Shahat in Libya) (273–192 BC) was a contemporary and correspondent of Archimedes. It is assumed that he received his education in Athens. One of his teachers was the poet Callimachus, who was curator at the library of Alexandria. Eratosthenes himself was later appointed chief librarian there by Ptolemy III. In 230, he was commissioned to educate Ptolemy's son Philopator.

Eratosthenes was an all-rounder, and his works include subjects from philosophy, mathematics, geography, astronomy, history and poetry. Unfortunately, none of those works had been preserved. Some of the titles are still known: "About Plato", "About the Comedy", "About the Survey of the Earth", "About the Constellation of the Stars". Other subjects touched on the calendar, Greek chronology since Troy and the chronicle of the Olympic Games.

His most important contribution was the advancement of scientific geography. And his single most important achievement was the determination of the radius of the Earth. There are various methods credited to him, with which he was supposed to have succeeded in this endeavour.

One of them was the explicit measurement of an angle to the sun with the help of an obelisk. This was done on the 21st of June at Syene (today Aswan in Egypt) at the summer solstice, when the sun was positioned perpendicularly. By measuring the angle of the zenith in Alexandria on the same day, the circumference of the Earth and thus its radius could be determined by taking the distance between Syene and Alexandria into account. This presupposes that both cities were situated on the same longitude, which was not exactly the case. The value thus obtained was 39,375 km, by today's standards a remarkably good value.

Other authors tell a different tale. Kleomedes describes the use of a hemisphere in conjunction with a sundial instead of direct measurement from an obelisk. Irrespective of the difference in method, one thing becomes quite clear: all Greek authors of the time were certain that the Earth was not a flat disk, as is often insinuated even in today's literature, but definitively a sphere. Some of the geographic results obtained then still served as a basis for Columbus' travel plans to India. Unfortunately he based his calculations not on the values of Eratosthenes but on those obtained by Poseidon of Apameis about 150 years later, which contained a much larger error. This value of the Earth's circumference had been passed on by Aristotle and was later cited in "Imago Mundi" by Pierre d'Ailly, who wrongly cited Roger Bacon from his Opus Majus, claiming that the distance between Spain and the western coast of India was small and to be sailed within a few days.

Ptolemy

Ptolemy lived during the second century AD. The astronomical and other scientific texts relevant at the time were all from the great Greek philosophers, foremost Aristotle. As an alternative to Aristotle, one could have resorted to the fundamental work of Hipparchus and Apollonius of Perga, who, together with the majority of their contemporaries, rejected the heliocentric world model of Aristarchus of Samos and Seleucus of Seleucia.

Ptolemy's famous contemporaries included among others the historian Cornelius Tacitus (55–117), author of the "Annals", "Histories" and "Germania," and the historian and biographer Suetonius Tranquillus (70–146). The emperor Trajan, under whom the Roman Empire achieved its largest extension (at about 100), reigned from 97 to 117. His adopted son Hadrian succeeded him as emperor (117–138). The most important ancient physician besides Hippokrates, Claudius Galenus, lived between 129 and 199. In Asia Minor, the prophet Montanus arose, to be rejected by Pope Hippolyt. In 165 Ireneus, bishop of Lyon, set up the canon of the Holy Scriptures. And finally, emperor Antonius Pius reigned from 138 to 161.

Claudius Ptolemy was born around 100 in Ptolomais Hermii in the Thebais in Egypt. He died in 175, probably in Alexandria. He was a Greek or at least Hellenized Egyptian. The name of Ptolemy appears quite often, referring to Macedonian rulers and is derived from the Ptolemaic dynasty in Egypt, although it is doubted that Claudius Ptolemy was related to the former Ptolemaic dynasty. His first name indicates the possession of Roman Civic Rights, which his family might have obtained during the time of the Claudian emperors, during the first half of the first century. Ptolemy was a Greek mathematician, geographer, astronomer, astrologer, theoretician of music and philosopher and spent most of his life in Alexandria in the vicinity of the research facilities and libraries there. Thus, he occupied the post of librarian at the famous library of Alexandria.

He wrote the "Mathematike Syntaxis (mathematical compilation), later called "Megiste Syntaxis" (major compilation), today known as the "Almagest", derived from the Arabic "al-magis", a treatise about mathematics and astronomy, covering thirteen books. The Almagest is the only completely preserved astronomical work of antiquity. It served as a standard work of astronomy until the end of the Middle Ages, was translated into Arabic and Latin and has been commentated abundantly. Besides an extensive catalogue of stars, it contains a multitude of astronomical data and Ptolemy's theory about the sun, the moon and the planets. The result is a refinement of the geocentric world model proposed by Hipparchus of Nicaea, later called the Ptolemaic World Model. With this, he, together with the majority of his contemporaries, did away with the heliocentric world model proposed by Aristarchus of Samos and Seleucus of Seleucia, which regained recognition only some 1300 years later in Europe through the works of Nicolas Copernicus, Johannes Kepler and Galileo Galilei.

To calculate the planetary positions, Ptolemy later published the "Procheiroi kanones" (manageable instruction charts). An enhanced theory of the planets was published under "Hypotheseis ton planomenon" (theory of the planets) later.

In the years after the Almagest, Ptolemy developed his geography, which consisted of a theoretical part, a descriptive part about Europe, Africa and Asia and a corpus of charts. This geography comprises the complete geographical knowledge of antiquity including cartography. It remained relevant until the age of discoveries.

During the years 147 and 148, he produced the so called Kanobis script, an inscription on a pillar in Kanobis during the reign of Antonius Pius. It contains improved astronomical data with respect to the Almagest.

Also widely distributed was his astrological treatise "Tetrabiblos" (the four books). From this, there exist several Greek manuscripts as well as Arabic, German and English translations. Among his scientific works, his contributions to optics and harmony in music theory bear mentioning, as well as a script about stereoscopic projections and an epistemological work called "Kriterion".

The geocentric world model places the Earth at the centre of the universe. Moon, sun and planets orbit the Earth geometrically along curved paths. According to the epicycle theory, the Earth remains at rest at the centre, but the planets no longer follow a perfectly circular route. Since Aristotle, it was generally assumed that the Earth had the shape of a sphere. This was the result of systematic observations over a long period of time and exact calculations. Until its replacement during the Renaissance, this world model was regarded as scientifically correct.

The geocentric world model was introduced during classical antiquity and prevailed, for example, over early suggestions by Aristarchus of Samos that not the Earth but the sun was at the centre of the cosmos. The geocentric world model was widely used in Europe. It was also taught in ancient China and in the Islamic world. It is not certain whether it had been common before the Greek domination in Mesopotamia. But Apollonius of Perge and Hipparchus had already employed eccentrics and epicycles to describe the planetary movements in their models.

Ptolemy worked with balance points around fictive points as well as with eccentrics and epicycles. Some people believe that Herakleides Pontikos developed a system in which the planets Mercury and Venus circle the sun, which together with the moon and the fixed starts, circles the Earth, which remains in its central position. This represented a compromise between the geocentric and the heliocentric world models. Tycho Brahe later proposed a modified version of it (s. further down in this chapter). The most important justification for the adoption of the geocentric world model was the observation of gravity, which could be explained by the fact that anything heavy would want to migrate to its natural place, the centre of the world. Aristotle himself had been an influential defender of the geocentric world model. But Aristotelian physics does not really get along with such hypothetical suppositions as eccentrics, epicycles and balance points.

Claudius Ptolemy used the so-called epicycle theory in what was called the Ptolemaic world model (Fig. 7.1). One challenge for the geocentric model had been the apparent backwards movement of the outer planets, for example, by Jupiter, against the background of the stars. This leads, from an earthly perspective, to an apparent

Fig. 7.1 The epicycle theory

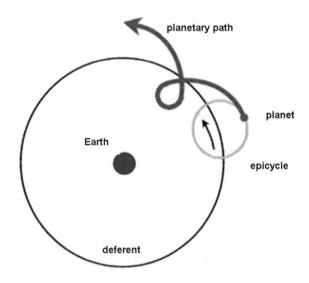

looping movement of the planet. This phenomenon, also called "retrograde", appears just when the planet is closest to the Earth. To harmonise the astronomical observations with the geocentric world model, it was necessary to let some of the heavenly bodies move along further circles around their original orbits. These are the so-called epicycles. According to this idea, the outer planets circle around an imagined point, which in turn circles the Earth. A planet initially moves along a uniform circle called deferent. Around this a second circle rotates, the epicycle. The planet itself moves around the epicycle in a uniform fashion. In this way the observed planetary revolutions appear as superposition of these movements. For some of these circles, additional tracks around these circles had to be modelled. Calculations within this model were extremely complicated. By employing about 80 such tracks, Ptolemy was able to harmonise the observations of planetary movements that were possible at his time with the geocentric model. For the sun, there was no retrogression. Ptolemaic astronomy combines planetary movements with the rotation of the sun under the assumption of geocentrism and succeeded within this complex model in producing largely correct forecasts.

Using the language of today's mathematics, one could call Ptolemy's calculation method a predecessor of Fourier analysis, with which the secondary periods of the planetary tracks were determined empirically. The Ptolemaic world model was far superior to the simpler, later, heliocentric model of Copernicus from the sixteenth century concerning forecasts of the planetary positions. Only after Kepler discovered that the planets were moving along ellipses did the Copernican model became sufficiently precise for the standards of the time and generally acceptable to astronomers. Ptolemy's calculation methods were extremely precise, much more precise than Kepler's for a long time to come, and his basic idea as a calculation method was also correct.

- "Our observations start with a view directed to the overall relation of the whole Earth to the whole sky. And furthermore, we will account for the ecliptic and about these locations in the part of the Earth, which is inhabited by us, and again concerning the differences between them regarding their respective inclinations of their respective horizons.
- As soon as this theory is understood, the rest will be much simpler. Thereafter we will report about the movements of the sun and the moon and special events in this context. Because, without having understood these correctly it is not possible to investigate the circumstances concerning stars with some benefit. At the end, this report will deliver an account about the stars. Since these things have to do with the sphere on which the fixed stars are situated, it is reasonable to have a look at those first and then at the ones which have to do with the five planets. And we will try to discuss all these things by using all origins and foundations concerning those we want to find out about, on the basis of the evident and certain phenomena observed by the ancients and by ourselves, and by demonstrating the consequences of these concepts by geometrical proof.
- And thus we have to state very generally that the skies are spherical and rotate spherically, that the shape of the Earth is quite reasonably spherical; concerning its position it is situated at the centre of the skies just like a geometrical centre; regarding its size and relationship to the fixed stars it behaves like a spot with no proper motion by itself. All these points will be dealt with successively and be brought to mind for us briefly" [8].

Of course, it is not possible to present the whole Almagest here, but only that section most interesting to out subject. Our interest is the postulation of the geocentric world model. The yardstick is the impact it had for more than one and a half thousand years on occidental thinking.

The section above appears between a general methodical-philosophical intro-duction and a detailed mathematical chapter, in which inclinations and planetary positions are calculated to support the theory. And indeed, they served well for many centuries to come. The question that imposes itself is of course: why discard a model that corresponds well with observations (without regard to any theological or ide-ological standards)? Other observations that invalidated the geocentric model were possible only much later (Galileo, Kepler).

Is it not because of purely aesthetical grounds that the complexity of epicycles was called into question? Or were there irrefutable rational justifications for objection then? Perhaps the question should be asked the other way round: why was the Earth placed at the centre in the first place? If this is because of religious or mythological reasons, then the question as to why these reasons justify placing the Earth should be allowed. Was this because of the importance man attached to himself, so that he must be at the centre of the world, or had this to do with daily observations? Aspects of these questions have been dealt with in Chap. 1 in the discussion of the anthropic principle.

The scientific context quite often leads to situations that seem to contradict solid experimental and theoretical findings: the theory of relativity, quantum mechanics etc. The path of the sun as observed by ancient people suggested that it was rotating around the Earth just like the moon (the latter fact remains part of our present day models). Obviously, for more than 1,500 years there was no a priori reason to question the geocentric world model. Sailors all based there navigation on it. The more important deviations were only determined by later observations of the sky.

Copernicus

Copernicus lived during the transition from the fading Middle Ages to modern times, and these times were indeed eventful ones, bringing about an abundance of changes and extraordinary personalities in politics, science and the arts. At his birth, the wars between the Houses of York and Lancaster over the crown of England were still raging, decided finally by Edward IV in 1483. Shortly after Copernicus' birth, the Burgundy Wars started, ending with the fall of the House of Burgundy in 1477. In Germany Maximilian I reigned over the Holy Roman Empire from 1493 to 1519. His epithet "The Last Knight" symbolised the doom of knighthood. He endowed the "Eternal Public Peace" in 1495. In England Henry VII, the founder of the Tudor Dynasty, reigned from 1457 to 1509, whereas his successor, Henry VIII, broke with the Catholic Church in 1533. In 1527 the Italian politician Niccolo Machiavelli died. As did Jacob Fugger two years earlier, under whom the Fugger family rose to become one of the most important trading dynasties in Europe.

During the lifetime of the astronomer, a number of discoveries took place, unsettling the world view in Europe. It all started with Henry the Navigator (1394–1460), a Portuguese Infante, who laid the cornerstone of the Portuguese sea and colonial power, followed by his compatriot Bartolommeo Diaz, who as the first European reached the southernmost point of Africa in 1487. And of course: Christopher Columbus reached the island of Guanahani, part of the Bahamas, on October 12th 1492, and thus discovered America. Finally, in 1498 Vasco da Gama reached India. This journey initiated the dominance of the Portuguese in the West Indian Ocean and started to divert important flows of goods and finances onto new trade routes, boosting Lisbon and Antwerp as world centers of trade and creating the preconditions for direct access to Asiatic goods by Europeans. In 1519, the first meeting between Hernan Cortes and Moctezuma II took place. One year later, Cortes defeated the Aztec forces under their leader Cuauhtemoc in the battle of Technochtitlan and destroyed the town.

These times of change saw the rise of great intellects that shaped occidental thinking for centuries to come. One of these paragons was Nikolaus Cusanus, universal scholar and philosopher, who lived from 1401 to 1464 and postulated the infinity of the universe for the first time. Martin Luther's (1483–1546) reformation revolutionized in the end the whole European continent. This epochal event had been prepared for by among others the humanist Erasmus of Rotterdam, who died in 1536. Another contemporary was the physician, alchemist, mystic, lay theologian and philosopher Paracelsus.

The onset of the Modern Age was also linked with names of important artists. Among the first was Filippo Brunelleschi (1377–1446), who invented the Central Perspective. He was followed by Leonardo da Vinci (1452–1519), the Italian inventor and painter, Matthias Grunewald (1470/80–1528), the German painter and master builder, his contemporary and compatriot, the painter and graphic artist Albrecht Duerer (1471–1528), and finally Michelangelo Buonarotti (1475–1564).

The time was equally rich in inventions. Johannes Gutenberg invented the set of movable letters and thus revolutionised the art of printing. For the first time, mechanical clocks appeared. It was altogether a time in which novel things and ideas were emerging, with which rulers and society had to come to terms: peacefully or by force. And into these times Copernicus published his findings.

Copernicus was born into a wealthy trading family in 1473 in Torun in Poland. After the death of his father, he was fostered by his uncle, the bishop of Watzenrode, who initially sent him to the University of Cracow. Later on, he studied in Italy at the Universities of Bologna, Padua and Ferrara. His subjects were law and medicine, but his personal university calendar at the University of Rome in 1501 already indicated an interest in astronomy. After having returned to Poland, he spent the rest of his life as a capitulary under his uncle, although he still found time to practise medicine and write about currency reform—and, of course to create his astronomical opus.

In the year 1514 Copernicus launched an extract of his thesis about the planetary movements into private circulation, but the official publication of "De revolutionibus orbium coelestium (About the Circulation of the Celestial Spheres)" with all mathematical proofs appeared only in 1543, after his impatient follower Rheticus had already published a short description of the Copernican system in the year 1541 on his own account. The "revolutionibus" demands a lot from the modern reader, since the mathematical methods of proof of the sixteenth century appear to be rather strange to our way of reasoning. The preface by Andreas Osiander, who was responsible for the printing, had not been authorised by Copernicus. Osiander inserted it without the knowledge of the author, and did not identify himself as the proper author. It states that the system presented was a purely mathematical hypothesis and did not represent reality. It was perhaps this preface that limited the initial debates about whether the work of Copernicus should be classified as heretical.

The fact that Nicolas Copernicus delayed the publication of "De revolutionibus" until shortly before his death can be interpreted as a sign that he was well aware of the uproar it would cause; his own preface, dedicated to Pope Paul III, anticipated many of the coming contestations. But he could definitively not have anticipated that later he would be counted among the most famous men of all time based on a book that would be read by only very few people and understood by even fewer.

The heliocentric world model is based on the assumption that the planets circulate around the sun. When the heliocentric world model was developed, it was an attempt to describe the structure of the universe.

However, at the time of Copernicus and even later, this world model presented a basic problem to most scholars: they assumed that, if the Earth were moving around the sun, then people and objects would fall down in a skewed fashion or even be catapulted into space; an object falling from a tower would reach the surface further west. The answer to this objection demanded a much better understanding of physics than was available. To blunt the attack made on theological arguments, the notion of the "hypothesis" was introduced. This was a kind of trick, suggesting that the entire thing was simply a kind of mathematical amusement, a playful example of calculation. The debates of the following one hundred years after Copernicus can only be understood in view of the fact that modern scientific thinking did yet not exist. In later epochs, an idea would either be confirmed or refuted by experiment.

Further, in considering the Copernican debate, one must take into account that at that time not all of the planets of our solar system were known, but only Mercury, Venus, Earth, Mars, Jupiter and Saturn. Another problem was the fact that the new heliocentric world model did not reflect the observed orbits of the planets as exactly as the geocentric one. The data basis was not sufficiently accurate.

"De revolutionibus orbium celestium" is much too long to be treated in its entirety here. For this reason, we regard only a short extract as an introduction and thereafter a section of the important Chapter 10. After a general preface, Copernicus postulates and then substantiates that the universe is spherical. After this, he presents a justification as to why the Earth should also be a sphere. He then proceeds to the derivation that the celestial bodies move in a uniform, eternal and circular way. He then asks what the consequences of these circular movements for the Earth are. After this, he relates the size of the universe to the size of the Earth. He explains why in antiquity people believed that the Earth was the centre of the universe. Finally, he refutes these old arguments.

This is the decisive extract from Chapter 10 [9]:

"Hence I feel no shame in asserting that this whole region engirdled by the moon, and the centre of the Earth, traverse this grand circle amid the rest of the planets in an annual revolution around the sun. Near the sun is the centre of the universe. Moreover, since the sun remains stationary, whatever appears as a motion of the sun is really due rather to the motion of the Earth. In comparison with any other spheres of the planets, the distance from the Earth to the sun has a magnitude that is quite appreciable in proportion to those dimensions. But the size of the universe is so great that the distance Earth-sun is imperceptible in relation to the sphere of the fixed stars. This should be admitted, I believe, in preference to perplexing the mind with an almost infinite multitude of spheres, as must be done by those who kept the Earth in the middle of the universe. On the contrary, we should rather heed the wisdom of nature. Just as it especially avoids producing anything superfluous or useless, so it frequently prefers to endow a single thing with many effects.

All these statements are difficult and almost inconceivable, being of course opposed to the beliefs of many people. Yet, as we proceed, with God's help I shall make them clearer than sunlight, at any rate to those who are not unacquainted with the science of astronomy. Consequently, with the first principle remaining intact, for nobody will propound a more suitable principle than that the size of the spheres is measured by the length of the time, the order of the spheres is the following, beginning with the highest (Fig. 7.2).

The first and the highest of all is the sphere of the fixed stars, which contains itself and everything, and is therefore immovable. It is unquestionably the place of

Fig. 7.2 The heliocentric
world model [9] according to
Copernicus and Galileo

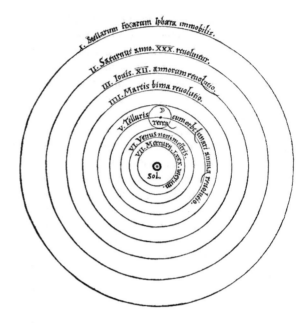

the universe, to which the motion and position of all the other heavenly bodies are
compared. Some people think that it also shifts in some way. A different explanation
of why this appears to be so will be adduced in my discussion of the Earth's motion
(I, 11).

(The sphere of the fixed stars) is followed by the first of the planets, Saturn, which
completes its circuit in 30 years. After Saturn, Jupiter accomplishes its revolution
in 12 years. Then Mars revolves in 2 years. The annual revolution takes the series'
fourth place, which contains the Earth, as I said (earlier in I, 10), together with the
lunar sphere as an epicycle. In the fifth place Venus returns in 9 months. Lastly, the
sixth place is held by Mercury, which revolves in a period of 80 days.

At rest, however, in the middle of everything is the sun. For in this most beautiful
temple, who would place this lamp in another or better position than that from which it
can light up the whole thing at the same time? For, the sun is not inappropriately called
by some people the lantern of the universe, its mind by others, and its ruler by still
others. (Hermes) the Thrice Greatest labels it a visible god, and Sophocles' Electra,
the all-seeing. Thus indeed, as though seated on a royal throne, the sun governs
the family of planets revolving around it. Moreover, the Earth is not deprived of the
moon's attendance. On the contrary, as Aristotle says in a work on animals, the moon
has the closest kinship with the Earth. Meanwhile the Earth has intercourse with the
sun, and is impregnated for its yearly parturition.

In this arrangement, therefore, we discover a marvellous symmetry of the universe,
and an established harmonious linkage between the motion of the spheres and their
size, such as can be found in no other way. For this permits a not inattentive student to
perceive why the forward and backward arcs appear greater in Jupiter than in Saturn

and smaller than in Mars, and on the other hand greater in Venus than in Mercury. This reversal in direction appears more frequently in Saturn than in Jupiter, and also more rarely in Mars and Venus than in Mercury. Moreover, when Saturn, Jupiter, and Mars rise at sunset, they are nearer to the Earth than when they set in the evening or appear at a later hour. But Mars in particular, when it shines all night, seems to equal Jupiter in size, being distinguished only by its reddish colour. Yet in the other configurations it is found barely among the stars of the second magnitude, being recognized by those who track it with assiduous observations. All these phenomena proceed from the same cause, which is in the Earth's motion.

Yet none of these phenomena appears in the fixed stars. This proves their immense height, which makes even the sphere of the annual motion, or its reflection, vanish from before our eyes. For, every visible object has some measure of distance beyond which it is no longer seen, as is demonstrated in optics. From Saturn, the highest of the planets, to the sphere of the fixed stars there is an additional gap of the largest size. This is shown by the twinkling lights of the stars. By this token in particular they are distinguished from the planets, for there had to be a very great difference between what moves and what does not move. So vast, without any question, is the divine handiwork of the most excellent Almighty."

Copernicus remains within the theological frame of reference of his time. For him, the questions of meaning or of the correct yardstick do not arise. His decisive scientific deed, which he justifies with the simplicity of his model, is the centralisation of the sun—but not only within our planetary system but as the centre of the entire universe. His references to the divineness of this celestial body are remarkable. Through this continuing centralisation, everything remains as it was except for his celestial mechanics: man is still in or near the centre of everything.

Of course, it was foreseeable that people who had grown up with the fixed belief that the Earth was at the center of the universe would find it very difficult to follow Copernicus' conclusions. Imagine this scenario: someone would come along today with the explanation the Earth is actually a cube, and even if all his calculations proved right, we would still have mental difficulties agreeing unequivocally with that.

Copernicus argues on the basis of observations and calculations, according to which the apparent backwards movement of our neighbouring planets are caused only by the movement of the Earth itself. This makes the resort to epicycles superfluous, which are indeed indispensable for justifying the geocentric model.

The limits of his model become apparent in two preconditions, which we know are of no use to us today:

- The spheres,
- The behaviour of the fixed stars.

Even Kepler continued sticking to the spheres. The notion of spherical, transparent, glass-like vaults, on which celestial bodies wander about, comes from pre-Christian antiquity. Indeed such a model seems to have been accommodated by heliocentricity, since with it, the spheres could be interpreted as ideal shapes of globes without the distracting epicycles.

With respect to the fixed stars, Copernicus of course is in error. We could not really expect differently because of the lack of observational instruments at the time. On the other hand, this also demonstrates that major parts of the old world model were saved in the new. Another interesting point is that Copernicus did not ask questions about the causes of the planetary movements. They were accepted as a given fact. It took Newton to create the mathematical apparatus with which the forces of the heavens could be calculated in relation to the movements observed. Copernicus' calculations were purely geometrical and with their help he detected the approximate correct geometry of the observed orbits—the circle appropriate for corresponding with the observed orbital periods. The mathematical foundations of his proof consisted of circular geometry, as well as angular mechanics for the determination of the planetary positions. He describes this method in detail in other parts of his work.

Tycho Brahe

Tycho Brahe proposed a world model that was a compromise between the heliocentric and the geocentric ones. During his lifetime, he was held to be the "Prince of Mathematicians"—of all time! His model, however, was rejected by Kepler. It did not play any significant role in science, but rather served ideological purposes. After Newton, it disappeared from astronomy completely.

Brahe was born on December 24th in 1546 (Julian calendar) in Knudstrup in Denmark. Already early in his life, he became fascinated by astronomy. On November 11th in 1572, he detected a new star in the Cassiopeia constellation, which made him famous. King Frederic II of Denmark provided him with the island Hven in the Oresund with 40 leaseholders on it, whose leases served as income for Brahe. It was here that he built his famous observatory "Uraniborg" in 1576, which became the centre of astronomical research. Brahe carried out his observations for more than twenty years and collected invaluable data. He also constructed a second observatory in Stjerneborg.

However, he fell into disgrace and had to give up Uraniborg because of ill-treatment of the leaseholders. He came under the roof of Count Rantzau in Wandsbeck, where he published his "Mechanic of the Renewed Astronomy" in 1598. During that time Brahe first came into contact with Johannes Kepler.

In 1599 Tycho Brahe was appointed Imperial Mathematician by Emperor Rudolph II, who provided the castle Benatek, southeast of Prague, as Brahe's place of work. His salary amounted to 3000 Guilders per years, an extraordinary sum at that time, to allow him to modify Benatek along the lines of Uraniborg.

Brahe went his own way with respect to Copernicus. He accepted the Copernican world model only partially by keeping the central position of the Earth just like Ptolemy, with the sun orbiting it, but at the same time having the planets orbit the sun. This was a compromise between Ptolemy and Copernicus.

Brahe engaged Kepler and gave him the task of compiling the Martian data from Brahe's observations. However, the relationship between Brahe and Kepler turned

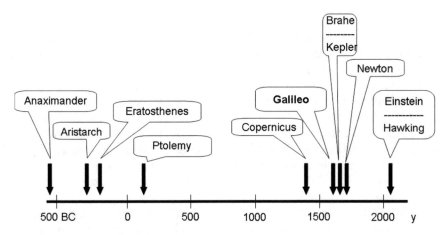

Fig. 7.3 Galileo's relative position in the Contest of Competing World Models

sour (s. section on Kepler further down), which lead to Kepler leaving the vicinity of Prague after various quarrels with his master, but returning later to attend Tycho Brahe on his death bed on October 24th 1601 in Prague. Brahe's main legacy was 24 folios containing data from many years of observations. These were the famous "Tabulae Rudolphinae", which later served as the basis for Kepler's own calculations of his own world model.

Galileo

Figure 7.3 shows Galileo's relative position on the historical timeline of the war of worlds.

Kepler

Kepler remained for a long time the only authority among his peers to have acknowledged the validity of the Copernican system. Of course, he had been familiar with Copernicus' publication for some time, but the intellectual roots of his reflections reached further back—to Nikolaus Cusanus, who lived from 1401 to 1464, and who had developed a type of geometrical mysticism. Another source, which in the end proved decisive for the success of Kepler's efforts, was indeed the existence of the observatory in Uraniborg, constructed by Tycho Brahe in 1576. Here the data for the "Tabulae Rudophinae", used later by Kepler, were collected.

Kepler was a contemporary of Galileo, and the technical invention of the so-called Galilean Telescope at that time, later improved by Kepler himself, paid off handsomely. Galileo was born before Kepler and died after him. So Galileo had been influenced by the same historical upheavals as Kepler, although some of the events had different impacts, since the situation in Italy was not the same as in the ruptured middle of Europe. Here a short summary:

The witch hunts in Europe had started. Shortly after Kepler's birth the French Wars on Religion between the Catholics and the Huguenots escalated with Bartholomew's Night in 1572. The Armada was defeated in 1588. And the colonisation of Canada by the French had commenced. During Kepler's lifetime, New Amsterdam was founded at the mouth of the Hudson River in 1624. At the same time the Thirty Years War, the end of which Kepler did not live to see, raged from 1618 to 1648. Other great contemporary intellects included the poets William Shakespeare and Miguel de Cervantes, both of whom died in 1616.

Johannes Kepler was born on December 27th 1571 in Weil der Stadt in Baden-Wurttemberg. Soon afterwards, his restless life, which led him to move from town to town until the end of his life commenced. During his youth, the unsteady business life of his father was responsible for his itinerancy; later, as an adult, the religious upheavals of the time were responsible for his wanderings. When he was five years old, his family relocated to Leonberg close by, where they stayed until 1579, moving at the beginning of 1580 to Ellmendingen and in 1583 back again to Leonberg.

Kepler was a fragile child and unsuited for bodily work, which prompted his mother to send him to the Latin School in Leonberg. There, Kepler passed the Stuttgart Country Exam in 1583, when he was not quite twelve years old. Because of his excellent results, he was allowed only one year later to enter the school of the erstwhile Premonstratensian monastery of Adelberg near the Hohenstaufen. Two years later, he switched to the erstwhile Cistercian monastery of Maulbronn, where he stayed for three years. Even as a young man, he felt the urge to mediate between the confessions. Not quite seventeen years old, he passed his baccalaureate in Tübingen, where he started his studies in the Tübingen monastery in 1589. At the age of nineteen and a half, he passed his magister exam. During his studies, he was interested in Platonic, Pythagorean and Euclidian writings. And he got to know the work of Copernicus, which, however, was discussed exclusively in private circles. Already then he became intrigued by the question of the distances between the six known planetary orbits (Mercury, Venus, Earth, Mars, Jupiter and Saturn), and he linked their spheres to the five Platonic Bodies: tetrahedron, cube, octahedron, dodecahedron and icosahedrons.

Initially Kepler wanted to finish his theological studies, but he followed a call to the monastic school in Graz to accept a post as mathematician. In Graz he taught logic, metaphysics, ethics, mathematics and astronomy. He attained a certain amount of fame through his "Prognostika"—astrological forecasts in the form of calendars. In Kepler's view, astronomy and astrology did not exclude one another. He believed that the so-called astrological "Aspects" could be derived from whole-number angular commensurabilities from the zodiac. Conjunction and opposition of celestial bodies are especially significant in this context. This fascination with the simple, age-old topic of whole-number proportions later became Kepler's starting point for his world harmony.

However, soon after Archduke Ferdinand took power in Styria, the Jesuit transition of the monastic school in Graz was started, in line with the Counter-Reformation. Initially Kepler and his family were spared, since they hoped that he would convert to the Catholic faith, but then, in 1600, he and his family had to leave town.

Previously, however, in 1597, Kepler had circulated his first work entitled "Vorläufer kosmografischer Abhandlungen, enthaltend das Weltgeheimnis über das wunderbare Verhältnis der Himmelskörper und über die angeborenen und eigentlichen Ursachen der Anzahl, der Größe und der periodischen Bewegungen der Himmelskörper, bewiesen durch die fünf regelmäßigen geometrischen Körper" (Forerunner of cosmographical tracts comprising the secret of the world regarding the wonderful relationship between celestial objects and about the hereditary and proper causes regarding the number and the periodic movements of celestial objects, proved by the five regular geometric bodies). He sent one copy to Galileo and requested a reply. In his response Galileo admitted that he was a follower of Copernicus. At that time Galileo was thirty-three years old. He had obtained the book and letter via the good services of Kepler's friend Paul Hamberger. Since Hamberger had to leave Padua the same day, and Hamberger had been charged to carry Galileo's reply back to Graz as soon as possible, Galileo's reply on August 4th 1597 was brief, courteously acknowledging Kepler's eminent status as a renowned scientist without delving into the subject matter of the book itself. He only had time to read the preface. However, besides confessing to be a follower of Copernicus, Galileo also explained why he had refrained from allowing to be recognized publicly as such. He cited the fate of Copernicus, who had by no means been persecuted by the authorities, but rather had been made the subject of ridicule and derision. Galileo expressed the opinion that as long as there was only a minority of scholars following the Copernican world model and a significant number of others still holding on to the old one, he was not prepared to go public himself.

In a letter about one month later to Michael Maestlin, Kepler's friend and patron, also a follower of the Copernican system, Kepler reported that the mathematician Galileo had been an adherent to the heliocentric system for many years and had requested two more copies of his work. He added a comment indicating that the work had been the combined product of himself and Maestlin.

Kepler then forwarded the requested additional copies to Galileo, accompanied by a letter of October 13th 1597. The largest part of the letter was taken up by Kepler's encouragement of Galileo to come out into the open and boldly defend the Copernican model in public, under the assumption that the prevalence of ignorance in Italy would be comparable to that in Germany. He did not agree with the idea of letting the unenlightened masses prevail over the truth and refused to resort to subterfuge to convince other scientists. At the end of the letter, he asked Galileo to carry out some observations of the Polar Star and Ursa Major in March 1598 since Kepler himself did not possess astronomical instruments of sufficient resolution.

At the same time, Kepler sent his "secret of the world" to the Dane Tycho Brahe, the "Prince of Mathematicians," known to be a fanatical adherent of exact observations, and then fifty years old. As previously mentioned, Brahe was serving Emperor Rudolph II in Prague and considered cooperation with Kepler. In the end, Kepler agreed. His first task was to recondition Brahe's Mars data without, however, gaining full access to the data collection. It came to conflict. In spite of this, the years in Prague until 1612 were Kepler's most fruitful. Only after Brahe's death in 1601 and on becoming his successor was Kepler able to dispose of Brahe's complete data collection. But before finishing his work on the data, he published a volume of four hundred and fifty pages in 1603, which dealt with the refraction of light. He needed four years to conceive his own solar system. In his work "Astronomia Nova" (New Astronomy), which was published in 1609, his first two planetary laws appeared. The farewell to circular orbits and the transition to ellipses were cause for some soul-searching for him. Other worries arose through Galileo's discovery of Jupiter's moons, upsetting the harmony of his world model. Later, however, through his own observations, he confirmed Galileo's observations.

Galileo himself had addressed Kepler again in April 1611 by using the good services of a certain Asdale in Prague asking Kepler's opinion of the "Dianoia astronomica" of Francesco Sizzi, an Italian astronomer who had first observed the movement of sunspots. In this treatise, Sizzi refuted on astrological grounds the existence of the Jovian satellites discovered by Galileo, the results of his observations having been published in "Siderius Nuncius". The subtitle of the Dianoia" continued with "…. where the rumor in the Siderius Nuncius be proved as being windy". Kepler compared Sizzi to Martin Horky, another sceptic of Galileo's discoveries. His judgement of Sizzi was very straightforward, comparing him to a blind man writing about the light of the sun. But in the end, he indicated that he wanted to avoid harsh criticism in public of the young author, and suggested befriending Sizzi to convince him.

After Emperor Rudolph's death, Kepler's position at the court in Prague was no longer assured, and he followed an invitation to Linz in 1612, where he stayed for a full fourteen years teaching mathematics and Latin. But in Linz he ran into problems because of his denomination and became isolated as a Calvinist. In 1615, he re-commenced his astronomical research. One of his self-imposed tasks concerned only a single detail, i.e. the calculation of the daily position of the planets, while his other occupations culminated in his major exposition on Copernican astronomy. He finished this in 1619 and called his opus "Joannis Kepleri Harmonices Mundi Libri Quinque" (Johannes Kepler's harmony of the world in five volumes). Here a short extract of its contents:

Volume 1: derivation of the notion of harmony on a geometrical basis

Volume 2: congruence of the regular polygons

Volume 3–5: dealing with music and the relationship between astrology and astronomy as well as mystical conceptions. His third Keplerian law can be found in Volume 5. Kepler had discovered it shortly before the Defenestration of Prague. It is remarkable that his three astronomical laws, which are still valid in physics today, were also made to serve to consolidate his mystical presuppositions.

There were further works. The work "Epitome Astronomiae Copernicanae" (the epitome of Copernican astronomy) was designed to serve as popular explanation of his world model. Furthermore, during his calculations, he came to appreciate Neper's logarithms, from which he then developed his own book of logarithms. And in the end in 1624, he completed the Rudolphinian Tables after twenty-two years of effort, comprising a systematic overview of the trajectories of all planets together with a catalogue of more than a thousand fixed stars, of which Brahe himself had already edited 777.

In the course of further successive uprisings, the situation in Linz became too hot for him and he moved on to Regensburg, and then further to Ulm, where he had the tables printed, only to return to Prague one year later. There he met Wallenstein, for whom he issued his famous horoscope. This was followed by Kepler's appointment as court astrologer in the town of Sagan in Lower Silesia. There he started another treatise about the moon, but left it unfinished. After Wallenstein fell into disgrace, Kepler got into financial difficulties as had been the case several times before, and he wanted to resolve outstanding business in Linz. His route took him via Regensburg on horseback, where he stopped at a trader's, with whom he was acquainted. Shortly after his arrival, Johannes Kepler fell ill with a high fever and died there on November 15th in 1630.

According to today's standards, the World Harmony could be labelled unscientific. In spite of that, the author had employed methods that were acknowledged as scientific at the time. And furthermore, he made fundamental discoveries, which can still be found in every physics textbook of today: his three laws about the movements of the planets.

Kepler's World Harmony covers three aspects:

- Planetary orbits or spheres.
- Geometrical bodies corresponding to these orbits.
- Musical intervals corresponding to the first two entities mentioned.

There is an additional relation: between geometrical bodies and the representation of the five elements of ancient Greece comprising:

- Fire for the tetrahedron,
- Air for the octahedron,
- Water for the icosahedrons,
- Earth for the cube and
- Ether for the dodecahedron.

As can be shown, each of these bodies possesses a surrounding sphere and an internal sphere. Kepler's endeavour now was to find a correct sequence of these bodies to match them with the planetary orbits. He tried to solve this problem using Euclidean geometry. His starting point was that the internal sphere serves as the innermost orbit for a planet relative to the orbit of the next planet, which moves around the external sphere of the first one: six planets and five bodies. To find his solution he had to deal with 120 possible permutations or arrangements. The best approximation for the relations between the planetary orbits resulted in the following sequence (Fig. 7.4).

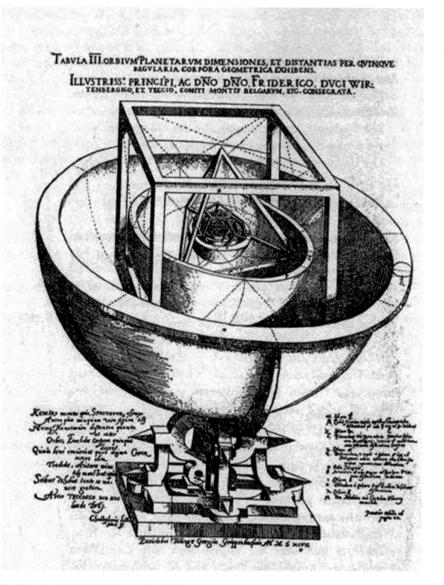

Fig. 7.4 Kepler's World Harmonics [10]

- The Earth as the internal sphere of the dodecahedron.
- Mars as the external sphere of the dodecahedron, which serves as the internal sphere of the tetrahedron at the same time.
- Jupiter on the external sphere of the tetrahedron, which serves as internal sphere of the cube.

- Saturn on the external sphere of the cube.
- The Earth at the same time on the external sphere of the icosahedrons.
- Venus on the internal sphere of the icosahedrons, which serves as the external sphere of the octahedron at the same time.
- Mercury on the internal sphere of the octahedron.

Now we come to music. Kepler sought the musical intervals, known already to Pythagoras, between the six planetary orbits. Again he had to tackle a multitude of permutations. In the end, he found two quantities satisfying his expectations: the maximum and minimum orbital velocities of the planets, i.e. the ratio of the speed at the perihelion nearest to the sun and the speed at the aphelion of the ellipse furthest from the sun. This resulted in the following values:

- Saturn: 5:4 or major third
- Jupiter: 6:5 or minor third
- Mars: 3:2 or fifth
- Earth: 16:15 or half-tone
- Venus: 25:24 or diesis
- Mercury: 12:5 or beyond the octave

With respect to the audibility of the music of the spheres, Kepler explained that it was limited in the sense that it could be captured only by the mind and not by the ear. In the end, Kepler allowed his quest for harmony to drive his research to such an extent that he abandoned any common ground for scientific consensus. Those who succeeded him would have to climb down from this height of harmony, at the cost of its nearly complete dissolution. Before we turn to the three laws of Kepler, here is a short diversion cited from the text [10]:

> Of course there are no notes in the sky and the movement is so strong that by the friction at the celestial air some sort of buzzing or whistling would occur. This would leave very little light. If this should communicate about the orbits of the planets it would either communicate it to the eyes or a similar sensory organ ….

Here we find two notions that have played an extraordinary role in modern physics: celestial air und light. If we replace celestial air with ether, we come very close to the Michelson experiment, which tried to prove the existence of the ether by measuring the speed of light. As is well known, this failed, and the result was that light propagates in all directions with the same speed—without regard to the direction or speed with which the system that incorporates the light source is moving. To explain this phenomenon, Einstein developed his Theory of Special Relativity.

Now let us have a look at the three Keplerian laws:

1. Planets move on ellipses with the sun residing in one of its focuses.
2. The running ray between the sun and a planet covers the same areas during the same time intervals (Fig. 7.5).
3. The square of the orbital periods of different planets relate as the third powers of the semi-majors of their elliptical orbits:

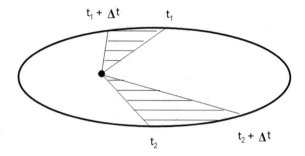

Fig. 7.5 Kepler's 1st and 2nd law

$$T_1{}^2/a_1{}^3 = T_2{}^2/a_2{}^3 = T_3{}^2/a_3{}^3 = \ldots.$$

Kepler's Laws are valid for all periodically recurring celestial bodies in our solar system, i.e. for the planets, but also for the Earth and its moon and of course also for artificial satellites around the Earth. Kepler's observation concerning his derivation of spherical harmony with respect to the maximum and minimum orbital speeds at the perihelion and aphelion is in accordance with his second law. This in turn is in accordance with the law of the conservation of energy. The potential energy is smallest at the perihelion, largest at the aphelion. Conversely, the kinetic energy and thus the velocity are highest at the perihelion and lowest at the aphelion.

The law of gravitation can be derived from Kepler's third law, with some simplification. This therefore takes into account the cause of the observed motion. Up until this time, at least, astronomical science had contented itself with only the description of the observed itself and had not looked for the underlying causes in a systematic way—with the exception of the mystical causes. This approach came to an end with Newton.

Newton

When Newton was born, the Thirty Years War was still raging. Furthermore, in the year of his birth, the English civil war between the crown and parliament began, which was decided by the parliamentary forces in the same year as the end of the Thirty Years War in 1648. King Charles I was beheaded, and England became a republic under Oliver Cromwell. Cromwell remained Lord Protector with absolute powers.

In science, Otto von Guericke in Magdeburg demonstrated the effect of the vacuum with his halved spheres. In England, especially in Cambridge, the Scholastic was still authoritative. New Amsterdam became New York, London was being ravaged the plague and major fires in 1665 and 1666. In 1669 the "Simplizissimus" by Grimmelshausen was published in Germany. Between 1672 and 1678, a new war, called the Dutch War, erupted in which England fought the Netherlands and Spain. Five years later, in 1683, the Turks beleaguered Vienna for the second time. In another revolution, William III of Orange toppled Jacob II and became King of England himself. The constitutional power of the monarchy was later secured by the Bill of Rights in 1689. There followed the Spanish War of Succession between 1701 and 1714. In the year of Newton's death Jonathan Swift published his famous novel "Gulliver's Travels".

Isaac Newton was born on January 4th 1643 in Woosthorp-by-Colsterworth in Lincolnshire in England and died on March 31st 1727 in Kensington. His father died prior to his birth. When his mother had married for the second time, Newton had to live with his grandmother for nine years until his stepfather died and he could return to his birthplace. He visited the elementary school in Grantham and at the age of eighteen went to Trinity College in Cambridge. Because of the great plague epidemic, the college had to close down and Newton had to return home after the conclusion of his studies in 1665. There he occupied himself for two years with the problems of optics, algebra and mechanics. During this time he was influenced by the writings of Descartes, Gassendi and Henry More. During these young years, he already became aware of important concepts that would later bear on his theoretical works on infinitesimal calculus, the nature of light and gravitation.

In 1667 Trinity College in Cambridge was re-opened and Newton returned there as a Fellow. As a precondition, he had to acknowledge Article 39 of the Church of England and deliver a vow of celibacy. He then could receive his clerical consecration after seven years.

Two years later, in 1669, he became holder of the Lucanian chair for mathematics there. In the same year Newton published a preliminary version of his infinitesimal calculus: "De Analysis per Aequationes Numeri Terminorum Infinitas". With this, he became one of the leading mathematicians of his time. Between 1670 and 1672, he taught optics with a special interest in light refraction. In this context, he constructed a reflecting telescope. In the same year, his publication "New Theory about Light and Colours" appeared in the Philosophical Transactions of the Royal Society. When his theories did not remain uncriticized, especially by his competitor Robert Hook, a leading personality in the Royal Society, the sensitive Newton withdrew from science for the time being and turned to alchemy. Then, in 1673, he started to become intensely interested in the Holy Scriptures and Patristic, an interest that he did not abandon until his death. As he rejected, in the course of his contemplations, the doctrine of the Holy Trinity, he demanded a dispensation from his obligation to receive clerical consecration. Clerical quarrels and the death of his mother drove him into a depression, which lasted until 1684. But already in 1679, he returned to his interest, the study of mechanics. This resulted in his treatise "De Motu Corporum" in 1684, which included some important conclusions that would be extended in the "Principia" in 1687. With these conclusions, Newton finally brought together and unified the findings from the motion experiments of Galileo, Kepler's observations of planetary movements and Descartes' deliberations about the problem of inertia. With his three laws of gravitation, he laid the groundwork for classical mechanics, which would enjoy uncontested validity until the early twentieth century. There were now no further obstacles to Newton's fame, even though Hook continued to accuse him of plagiarism of his own ideas.

There followed a succession of events and activities that had only little relevance to his scientific career: resistance against King Jacob II, who wanted to transform the university into a Catholic institution, further theological studies and—in this context—an exchange of letters with the philosopher John Locke, his role as delegate of the university to parliament; then again nervous problems. In the end—through the intervention of the Earl of Halifax—Newton became first Warden in 1696, then in 1699 Master of the Royal Mint in London. Since Newton took this office, to which lucrative sinecures were attached, rather seriously, his scientific activities had to retreat in the background. In 1699, he was awarded the high honour of appointment as one of only eight foreign members of the Academie Francaise in Paris. In 1701 he stepped down from his professorship in Cambridge, but published anonymously another law concerning the cooling of solid bodies in air. In 1703 he was appointed president of the Royal Society, an office which he held until his death. After Hook had died in 1704, Newton published his own work on optics. Thereafter, his dispute with Leibniz about the claim to discoveries in infinitesimal calculus commenced. Newton lived reclusively in his house in London, which harboured a small observatory, and dedicated his life entirely to the study of ancient history, theology and—how could it be otherwise—mystics. Newton died as a wealthy man, despite a massive speculation loss, and was buried in Westminster Abbey.

Before turning to gravitation, just a few remarks concerning Newton's optics and his theory of light:

In his treatise "Hypothesis of Light" Newton introduced the concept of ether and the particle nature of light: light particles move through a material medium—this was pure materialism. But the particle theory of light could not explain such phenomena as interference. Newton therefore soon got into a dispute with Huygens, a representative of the wave theory. The ether was abandoned two centuries later on account of the Michelson experiment, and wave and corpuscle are today described and unified by quantum theory. Furthermore, Newton's notions about space and time, both of which he took to be absolute, were put into a different perspective with the publication of the General Theory of Relativity.

As mentioned several times already, early authors contented themselves with the pure description of astronom ical observations and in so doing brought phenomena into accord with some sort of order. The real causes of planetary movements and cosmic appearances, especially with respect to their origin and future destiny, were not really sought after. Newton brought these approaches to an end. He described celestial phenomena as consequences of natural laws. These laws were able to describe the movements of the planets and satellites without fault.

To arrive at this new approach, Newton had to propose a thesis—a thesis, which we believe is still valid today for the whole of cosmology, i.e. that physical laws that have been proved on Earth apply equally everywhere else in the cosmos. In Newton's time, this was possibly not such a daring supposition as may appear to us today, since anthropocentrism as the cosmic point of reference still prevailed as a common intellectual good. Today, at a time when such shared implicit understanding a priori no longer exists, such a claim, although it may be necessary, is nevertheless daring. Under this assumption, Newton in this way reached a synthesis between gravitational effects already observed by Galileo and the best measured data provided by Kepler. Now a completely new vision for the world opened up, which for the first time was free from all mystagogical elements and would remain so up to the present day.

Kepler showed us in a way the climax of order, a perfect cosmic harmony. Newton dissolved it. Through the consequent application of the gravitational laws, he also took into account the mutual interaction of celestial bodies, resulting in many body problems, the exact solutions of which remain to this day possible only with numerical methods supported by powerful computers. The orbits of celestial objects become ideal trajectories under their mutual interactions, which in reality deviate because of perturbations. This reality now proves to be too complex to maintain the perfect order of the cosmos. There is still no chaos but the delimitation to it has become more difficult.

If we want to transform Newton's postulates into mathematical language, then we need an appropriate frame of reference for describing motions, which leads us to kinematics and the equations that describe such movements. In our normal world, the reference frame would be the Cartesian coordinate system. Then movement means the changing of the position of an object over time. The rate of this change is called velocity. If velocity also changes over time, then one talks of acceleration.

Kinetics now brings together force as the cause of motion and kinematical descriptions of motions. This is the synthesis of Galileo and Kepler. Movement only takes place if a force acts upon an object. Newton had observed that the alteration of motion is proportional to the force applied to a mass and the direction of the force. The proportionality of this behaviour is ensured by the mass itself.

Another discovery by Newton was that each force provokes the action of a counter force:

$$\text{actio} = \text{reactio}$$

Kinetically speaking this means that motion has to overcome resistance, usually understood as inertia. Newton's equations do apply to forces acting directly as well as at a distance.

In the event that two masses exercise forces upon each other, the law of gravitation is applicable: the gravitational constant multiplied by the two masses and divided by the square of the distance between them. We must differentiate between mass and weight. Weight is a property of mass and proportional to it. This property is determined by the attraction of the Earth and thus can be measured. The coefficient of proportionality is the acceleration of a free falling body, which is $g = 9.8067$ m/s^2. When calculating the drop height or the free fall velocity, in both cases the mass does not appear in the relevant equations!

When Newton derived his equations, he did not use an inclined plane but rather a pendulum.

Einstein

Albert Einstein was born on March 14th 1879 in Ulm. He died on April 18th 1955 in Princeton in the USA. Although his parents came from long established middle-class Jewish families in Swabia, his surroundings were rather assimilated and not very strongly religious. Shortly after his birth in 1880, the family relocated to Munich because of his father's business.

Of his childhood and youth, we know that he was rather a late bloomer. He started to speak only at the age of about three. But already as a five year-old, he became interested in playing the violin. After attending elementary school, in 1888 he went to the Luitpold High School, today Albert-Einstein High School. After his father's business closed down, the family moved to Milan in 1894. Initially Albert was to stay behind to obtain his diploma from secondary school. However, for various reasons he left high school at the age of fifteen without a diploma and followed his family. Two years later, he gave up his German citizenship to avoid military service, and at the same time he left his synagogue congregation. Through the good services of an acquaintance, Einstein was able to visit the Canton School in Aarau in Switzerland, where he obtained his matriculation standard in 1896 with distinction.

His father wanted him to study electrical engineering. But Einstein applied to study at the Zurich Polytechnic. After failing his entrance exam there in 1895, he later was admitted after a second try, after having gained his university entrance diploma. Einstein finished his first scientific work, entitled "Über die Untersuchung des Ätherzustandes im magnetischen Felde" (About investigations of ether states in a magnetic field) at the age of only sixteen. It was never published. He was not very keen on high school, especially with regard to mathematics. He later compensated for his deficits in this field by relying on the assistance of others.

Einstein finished his studies in the year 1900 with a teaching degree for mathematics and physics. Since he could not obtain a post as assistant professor at his institution, he started to work as a tutor in Winterthur, Schaffhausen, and later in Bern. In 1901 he obtained Swiss citizenship. In 1902 he got his first real employment: at the Swiss Patent Office in Bern as a technical expert 3rd category. After his marriage, Einstein continued to live with his first wife in Bern until 1905.

1905 became "his" year. At the age of 26, his work "Über einen die Erzeugung und Verwandlung des Lichts betreffenden heuristischen Gesichtspunkt zum photoelektrischen Effekt" (On a heuristic aspect with regard to the photo-electric effect concerning the generation and transformation of light) was published. For this essay, he would later win the Nobel Prize. During the same year, he handed in his dissertation about "Eine neue Bestimmung der Moleküldimensionen" (A new determination of molecular dimensions), to obtain his Ph.D. Scientific results now came one after another. One month after his dissertation, he published "Über die von der molekularkinetischen Theorie der Wärme geforderte Bewegung von in ruhenden Flüssigkeiten suspendierten Teilchen zur Brownschen Molekularbewegung" (About the movement required by the molecular theory of heat of particles suspended in a stationary liquid regarding Brown's molecular movement hypothesis). And finally, once again only one month later, he submitted the paper "Zur Elektrodynamik bewegter Körper" (the electrodynamics of moving bodies) with an appendix, in which for the first time the famous formula $E = mc^2$ appeared, to the Annalen der Physik. The article appeared on September 26th 1905. Already the next day Einstein delivered an appendix with the title: "Ist die Trägheit eines Körpers von seinem Energieinhalt abhängig?" (Does the inertia of a body depend on its energy contents?). Special Relativity had come to life.

Carl Friedrich von Weizsäcker wrote many years later about this year of marvels:

In 1905 an explosion of genius happened. Four publications on diverse subjects, each of them worthy of a Nobel Prize: Special Relativity, the hypothesis of light quanta, the confirmation of the molecular structure of matter, Brown's motion, the explanation of heat in solid bodies with the Quantum Theory."

After the rejection of Einstein's first habilitation application in 1907 at Bern University, he was appointed lecturer in theoretical physics at Zurich University. Two years later, in 1911, he obtained a one-year appointment as a regular professor for theoretical physics at the University of Prague, where he became an Austrian citizen. In 1912 he returned to the Polytechnic in Zurich as a professor, to the institution at which he had failed the entrance exam in 1895.

Max Planck finally asked Einstein in 1913 to come to Berlin as a full-time paid member of the Prussian Academy of Science and from 1914 onwards as director of the Kaiser Wilhelm Institute of Physics. Two years later, in 1915, he was able to publish the results of his further studies as his main legacy, the General Theory of Relativity. During his time in Berlin, he also came into contact with Max Wertheimer, the founder of Gestalt Theory, for which Einstein developed a special interest. Perhaps Wertheimer's ideas contributed indirectly to Einstein's later search for a unified field theory.

After the experimental confirmation in 1919 by Eddington of Einstein's theoretical prediction about the diversion of light by the sun's gravitation, even the non-scientific world paid attention to Albert Einstein. He was now on the way to becoming a legend, a status never given scientists before him and never again after him in quite this form. He then was awarded the Nobel Prize for physics "for his merits in theoretical physics, especially for his discovery of the laws of the photo-electric effect."

When he was awarded an honorary doctoral degree at Princeton University, he planned to teach in Princeton, New Jersey, for one half of each year and in Berlin for the other half of the year. After another stay in the USA, he did not return to Germany after the National Socialists came to power in 1933. This step was followed by a series of voluntary resignations from academies and societies in Germany and Italy, his abandonment of his German citizenship, as well as his ostracism by the German state apparatus. His contacts to Germany had ruptured. In the year 1940 he obtained American citizenship.

Earlier, in 1933, Einstein had become a member of the Institute for Advanced Studies, a private research institution in Princeton, the town in which he continued to live until his death. He consecrated his remaining years as a researcher to the quest for a unified field theory, i.e. the unification of gravitation and electromagnetism. He did not reach this objective—just as no one else up to the present has done so. During this time, the atomic bomb was developed. This and the scientific foundations for its creation had never been Einstein's subject of research. He had been urged by other researchers to bring his reputation to bear and call for the American President to accelerate the development of this weapon.

Einstein died on April 18th in 1955 at the age of 76 in Princeton through internal bleeding caused by the rupture of an aneurysm.

From time immemorial great intellects have studied free fall. The interesting question is of course: what would happen, if such experiments—simple free fall all by itself—were to be executed in different systems moving at different speeds, which need not be constant, with respect to each other. The systems being compared may even accelerate with respect to one another. In such an experiment, one system may even be the laboratory and the other is represented by the free falling object itself.

The answer is given be the General Theory of Relativity:

"If we assume that in two systems the same physical laws do apply, then there exists no reference system for absolute acceleration, just as there is no absolute speed in Special Relativity."

Or otherwise:

"Once physical laws are valid in one environment they are equally valid in an environment being in relative motion to it."

The precondition is the determination that:

"In every arbitrary local Lorentz reference frame, everywhere and at any time in the universe, all physical laws (with the exception of gravitation) adopt the well-known form of Special Relativity."

This sounds much like the proposition by Newton about the validity of physical laws everywhere at any time. Before looking into the consequences of this principle, we should clarify some of the concepts on which it is based. One of them is space time.

Space time can be experienced in daily life and connotes a four dimensional entity with the three space dimensions length, breadth and height with time as the fourth dimension.

Within space time all events in the world can be represented. An object is always present somewhere at any point in time. The change of position over time becomes a graph in four dimensions: a world line. This applies to a soccer ball, but also to a human being. Everything seems frozen in the space time matrix, because movement is already fixed as world line itself—everything forever and anywhere. One characteristic of this continuum is that all four coordinates are on an equal footing. This means for example that space and time coordinates can be exchanged under certain conditions.

Within space time, things can be described, for example the distance between two points. The shortest distance between two points is called a geodesic.

Another basic concept is that of a metric. A metric measures the distance between two points—in the space time continuum therefore between two events (time and space).

As was stated above, there exists no absolute reference system. To clarify this, let us assume that somebody is standing on the roof of a building and throws a ball downwards. In classical terms, one would say: free fall, straight line, shortest path. The geodesic would in this case be a straight line. But it is not as simple as that. There are more influences.

First of all, there is the Earth's rotation. Furthermore, the Earth is moving around the sun and so the ball follows an arc. But then the sun does not stay in one place either, but rotates within our Milky Way. A distant star sees another arc-like movement … and on and on with galactic clusters and the whole cosmos. The origin of the coordinate system of the universe is certainly not on the roof of that house. The accelerated ball describes a complex track on a geodesic, which may be the shortest path between two points in space time but is not a straight line.

One reason for the complexity of the description of this movement is the use of our Euclidian coordinate system. In the vicinity of a local environment, the Lorentz transformation applies. The four dimensions, including a time like one besides the three space dimensions, are called Minkowski Space. According to Einstein, physics is simple in this region, but it becomes complicated in global space.

Let us return to acceleration. Our notion of acceleration is derived from the fact that we rest on solid ground and see all sorts of objects falling down around us. At the same time, we forget that we ourselves on our world line are in accelerated motion which includes the ground on which we are standing. But we measure this process in flat Minkowski space, restricting us to our well known kinetic equations.

Let us imagine that we are in a spacecraft that contains a variety of other objects such as a bunch of keys, pieces of money, screws and even peas, in addition to us. This space craft accelerates through space. But all of the objects remain at rest: they follow a straight line. In spite of that, we do sense the effect of acceleration, when, for example, we are pressed back into our seats. Acceleration therefore is no illusion. Under the laws of Special Relativity, we now can dissect space time into small consecutive segments and connect them in sequence. In each of these segments, the laws of the Lorentz transformation are valid. But at the same time, we realise that for example time dilation and length contraction do change along the path of acceleration with the change of speed. The result at the end of this mathematical effort is that space and time follow a curve. And now we enter the realm of General Relativity, and it is time to change the coordinate system.

Instead of describing accelerated motion as a curve in flat space, one can invert the approach and in this way obtain a "natural" coordinate system: if an accelerated body moves on a geodesic, i.e. on a line representing the shortest distance between two points, then the space, in which this takes place, must inevitably be curved. If we move within a local inertial system freely along such a geodesic, then we will observe all free falling objects to move with constant velocity.

Global curved space is of course the home of many geodesics spanning space time in the first place. Once we have eliminated acceleration through the curvature of space time, we have abolished at the same time any force acting due to its mass. Or in other words: space time tells a mass how it has to travel. Inversely, mass tells space time how it has to curve. This is the rule of gravitation. How can it be expressed mathematically?

If space time can be expressed by a tensor (a mathematical entity consisting of rows and columns like a determinate or a vector), the Riemann Curvature Tensor, on one side of an equation, then its equivalent on the other side of the equation has to be a tensor as well.

Each event in four-dimensional space time carries with it a so-called stress-energy tensor that contains information about energy density and impulse. The final equation at which we arrive expresses that the curvature of space is proportional to the stress-energy present.

Hawking

The basis for our modern cosmological models is the General Theory of Relativity. Other contributions that shape the present-day model include the cosmological constant of Edwin Hubble, mentioned earlier, and the Schwarzschild solution of Einstein's field equation. It was published by Karl Schwarzschild only one month after Einstein's publication of his General Theory of Relativity and describes a gravitational field outside a spherical, non-rotating mass such as a star or a black hole. It is the most general, spherical and symmetrical solution of Einstein's field equation. A black hole according to Schwarzschild is surrounded by a virtual spherical surface called the event horizon and possesses the so-called Schwarzschild radius.

Stephen William Hawking was born on January 8th in 1942 in Oxford. He died on March 14th 2018 in Cambridge. At the time of his birth, his parents were living in oxford, but left for St. Albans near London in 1950, where Hawking attended school from 1953 onwards. It was initially planned that he should study medicine, but already prior to his graduation he had participated tentatively in an entrance exam at Oxford University, which he passed and thus obtained a stipend. After his bachelor's degree, he wished to study cosmology in Cambridge, but was lacking the relevant entrance exam there, so he volunteered for an oral exam, which he passed in 1962 with top results. In Cambridge at Trinity Hall, he received a PhD in theoretical astronomy and cosmology in 1966. There he became a research fellow and afterwards professorial fellow at the Gonville and Caius College. Already in Oxford first signs of his disease, amyotrophic lateral sclerosis, were noted, which then became worse during his time in Cambridge. This disease destroys the nervous system. Since 1968 he has been confined to a wheel chair.

Hawking worked with Roger Penrose, with whom he proved the existence of singularities in General Relativity. For this work, he received the Adams Prize of Cambridge University in 1966. Thereafter he moved to the Institute of Theoretical Astronomy, where he worked until 1973, to develop the theory of black holes at the Institute for Applied Mathematics and Theoretical Physics.

In his further published results, he continuously refined or revised his cosmology as well as details of black holes. These included:

- Annihilation of black holes
- Relationship to Quantum Field Theory
- Quantum gravitation
- Quantum cosmology
- Thoughts about open and closed universes.

Hawking made a name for himself as author of popular science books, among others "A Short History of Time". In addition, he dealt with questions about creation, the origin of the universe and God. As one consequence, he became a lifetime member of the Papal Academy of Science.

In the meantime, his illness had grown steadily worse. After a tracheal operation, he was rendered unable to speak. A special speech computer was developed for him. Later he could no longer operate the machine by hand. He worked with the aid of infrared communication achieved by means of a transmitter fixed to his spectacles and directed to his computer. The transmitter was controlled by movements of Hawking's cheek muscle.

The Standard Hot Big Bang Model (Fig. 7.6) is based on the fact that gravitation completely dominates the development of the universe, but that the observed details are still determined by the laws of thermodynamics, atomic physics, nuclear physics and high energy physics.

It is assumed that during the first second after the beginning, the temperature was so high that there existed a complete thermodynamic equilibrium between photons, neutrinos, electrons, positrons, neutrons, protons and various hyperons and mesons and possibly gravitons.

After several seconds the temperature dropped to about 10^{10} K, and the density amounted to about 10^5 [g/cm^3]. Particles and anti-particles had annihilated themselves, hyperons and mesons had decayed and neutrinos and gravitons had decoupled from matter. The universe consisted of free neutrinos and perhaps gravitons, the field quanta of gravitational waves.

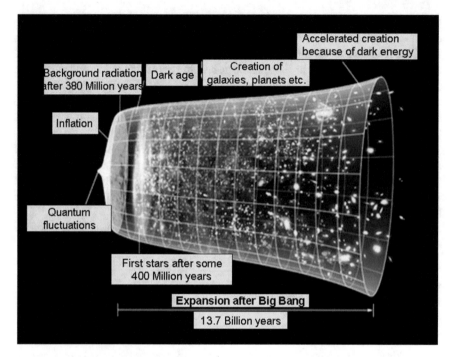

Fig. 7.6 The big bang model (© NASA modified)

During the following period between some 2 and 1000 s a first primordial creation of elements took place. Before this, any such occurrences were destroyed by high energy photons. These elements consisted primarily of alpha-particles (He–4), traces of deuterium, He–3 and Li. The rest were hydrogen nuclei (protons). All heavier elements originated later.

Between 1000 s and 10^5 years, the thermal equilibrium was maintained by a constant transfer of radiation into matter as well as by permanent ionisation processes and the building of atoms. Towards the end of this period, the temperature dropped to a few thousand degrees. From then on, the universe became dominated by matter instead of radiation. Photons no longer had enough energy to, for example, ionise all hydrogen atoms permanently.

When photon pressure had disappeared, condensation of matter into stars and galaxies could begin: between 10^8 and 10^9 years afterwards.

Hawking's cosmological standard model utilizes a mathematical and physical phenomenon called a singularity. A singularity is something like a defect in a coordinate system, like the function $y = 1/x$. Yet another singularity is for example the North Pole. All meridians meet at a single point. A singularity thus describes something unique. In the singularity postulated by Hawking, space time is concentrated in a single point. In the language of General Relativity such a singularity can be described as follows:

In a space time manifold one has to take into account:

- All space-dependent geodesics
- All null geodesics (photon trajectories)
- All time-dependent geodesics (free falling observer)
- All other time-dependent acceleration trajectories.

Assuming that one of these trajectories ends after a finite distance, and assuming that it is impossible to expand the space time manifold beyond this end point—then this end point is called a singularity.

The singularity upon which the model of the expanding universe is based, is generally known as the Big Bang. This means that there was a point in time at which the universe was infinitely small and infinitely dense. Under these circumstances, none of our laws of nature were applicable, and there was no way to predict the future. If there ever were events that took place prior to this point in time, then they would never have been able to influence things happening today. For this reason, any such events may be ignored, since they neither can be observed nor can they influence our present situation. On this basis, one can claim that time started with the Big Bang. Modern cosmology states that—whether chaos or some other state reigned, ordered under a different principle than chaos—in any case some kind of singularity had to be involved, the transit of which is not allowed, so that the question about any precedent state does not make sense.

Singularities and general aspects of relativity play not only an important role in the Big Bang, but also with respect to the life cycle of stars. Imagine a celestial body possessing already a certain density. According to the Schwarzschild model, which relies on hydrostatic laws, this density should be homogeneous with a spherical shape. Its central pressure depends on its radius. If this body is a star, there are two possible ways for its life cycle to end in a stable equilibrium:

- Offloading superfluous mass
- Collapse.

One of these two options will occur, once the nuclear reactions of the star cease because of the exhaustion of fusionable material, ending the fusion that has maintained stability until then. In our further consideration, we will consider only the option of collapse.

A collapsing star becomes a singularity once it surpasses a critical density. The corresponding radius for this threshold is called the Schwarzschild radius. For our Earth, it would be 9 mm. The singularity into which a star collapses is called a black hole.

Related to the Schwarzschild radius is the event horizon. It defines a boundary in space time that cannot be passed. This is true not only for other celestial bodies or bold travellers who want to pass this virtual membrane, but also for electromagnetic waves—i.e. light. Therefore its name: "black hole". The reason for the existence of this boundary is to be found in the enormous curvature of space time.

Does this mean that it is impossible to find black holes at all because they cannot be observed? Not necessarily. There are a number of indirect phenomena indicating the existence of black holes:

- Direct observation of the collapse of a star.
- Direct observation of the creation of many black holes, when star clusters consolidate and super massive galaxies originate.
- The growth of black holes dating from early times of the universe; these objects continue to attract other masses, which emit radiation as they disappear into the black hole.
- Observation of the spiral path of a celestial body when approaching a black hole.

Relativistic effects play a role in four more star configurations besides black holes:

- White dwarfs
- Neutron stars
- Super massive stars
- Relativistic star clusters.

Here we want to have a brief look at the first two of these configurations.

White dwarfs as well as neutron stars may come into being—just like black holes—at the end of the lifecycle of a star. What does the final stage in the life of a star like our sun look like?

Someday, the thermonuclear process that keeps the star burning will come to an end, because the necessary fuel is exhausted. This is the case when large quantities of Fe–56 and Ni–62 have been produced—those nuclei with the highest binding energy. What happens next depends on the mass of the star. A guideline is the so called Chandrasekhar value of about 1.44 times the mass of our sun. At 1.44 to 3 times the mass of the sun, the star will become a neutron star. If the star's mass is less than this value, then it will become a white dwarf. Beyond 3 solar masses, it will collapse into a black hole.

The temperature and thus the brightness of white dwarfs is determined entirely by a degenerated electron gas. In neutron stars the gravitational pressure is so large that electrons and protons recombine to neutrons. These constitute a supra liquid in the star's interior that has no internal friction.

It is tempting to risk a comparison between Hawking and Anaximander. Both theories are somewhat shrouded, for the layman, in mysterious language. Both seek to proclaim final truths. One question which arises, is: does one or the other perception influence our daily existence? Or the other way around: if Anaximander came not at the beginning but at the end of the whole cognitive process that we have discussed so far and Hawking at the beginning—what would be the difference for us today? Would there be any?

Chapter 8
Florence Revisited—Dialogo and Its Consequences (1624–1633)

Saint-Malo

The mouth of the river Rance near Saint-Malo on the northern shores of Brittany in France offers a magnificent sight—but not only because of the landscape, but also because of a technical construction that is without equal in the world. If one takes a trip by pleasure boat into the entrance of the water gate, initially nothing indicates the presence of this large-scale technical project. Once the movable bridge above the water gate is raised, traffic comes to a standstill on the mighty levee that connects Saint-Malo with Dinard in northern Brittany.

The secret of the technical machinery in question is not visible at first sight but is hidden below the surface: the Tidal Power Station at the Rance (Fig. 8.1). But even the impressive view into the hall inside the dam below the roadway does not necessarily reveal the purpose of the installation. However, there exists a multi-media exposition area of 300 m^2 about the tidal power station, which initiates the visitor free of charge into the secrets of energy transformation by tidal forces: the first tidal power station worldwide, constructed between 1962 and 1965. The major project on the northern shores of Brittany delivers 600 Million kWh per annum in the form of electrical energy and can thus supply on its own about 300 000 inhabitants with electricity.

The electricity is created by 24 mighty generators, each of which can produce a maximum power of 10 million Watts. They are driven by water currents induced by the tides twice a day in one direction (from the sea into the mouth of the Rance) and twice a day in the opposite direction (back to the sea), passing through the turbines: day after day.

The location of the tidal power station had been chosen on purpose. The tidal difference in the mouth of the Rance is among to the largest on Earth and can reach 14 m. The reason for this is the shallowness and the backwater effect of the channel.

In the meantime, other countries have followed the Rance example. Today there exist tidal power stations in Canada, China, Russia, South Korea and the UK.

© Springer International Publishing AG, part of Springer Nature 2018
W. W. Osterhage, *Galileo Galilei*, Springer Biographies,
https://doi.org/10.1007/978-3-319-91779-5_8

Fig. 8.1 Tidal power station near St. Malo (W. Meinhart, Hamburg)

Back to Florence

In 1624 Galileo travelled to Rome and was received by Pope Urban VIII (Fig. 8.2) for the second time, for six private audiences. The Pope even encouraged Galileo to concern himself with the theories of Copernicus—even to publish about them—under the pre-condition that he would treat these ideas purely as hypotheses. The Pope also praised Galileo in a letter to Archduke Ferdinand II of Tuscany, in which he stated that the scientist had gained his personal benevolence.

Galileo was in fragile health, like many of his contemporaries of similar ages; medical care as we know it today was lacking. He took heed of the advice of the Pope and started to grapple with the new world model in earnest and in a comprehensive way. After long and tedious preparation, he needed a full six years to complete his reckoning with Ptolemy and Aristotle. His approach was subtle enough, since he shrouded it into a dialog (which was more like a conversation among three, since it involved three fictitious persons; the word dialog probably referred to the two world models that competed with each other) and left the conclusion to be judged by the reader rather than committing himself by name and in writing. However, to the authorities it was clear enough that the tract constituted a full defence of Copernicus by Galileo, and from there on things took their course.

Fig. 8.2 Gian Lorenzo Bernini: Pope Urban VIII between 1635–1640, Musei Capitolini, Rome

Unfortunately for the reader of today, the main pillar of Galileo's justification for the Copernican system was his explanation of the tides as caused by the Earth's motion, which was scientifically wrong (more of this later at its proper place)

He entitled the result of his endeavours the "Dialogo di Galileo Galilei sopra i due Massimi Sistemi del Mondo Tolemaico e Copernicano", translated: "Dialogue by Galileo Galilei about the two most important world systems, the Ptolemaic and the Copernican", published in 1632. For all practical purposes, the work could have been entitled "Dialogue about the two most important systems in natural science, the Aristotelian one and the observed one" or something similar.

During the preceding sixteen years, i.e. after the first rebuke in 1616, some of his pupils had already started to apply his findings, on which Newton founded his laws of motion half a century later, in practise. Then the Dialogo appeared. This work had an immense impact.

Initially Pope Urban VIII indulged himself with the illusion he could prove the affinity of the church to the sciences by using this publication as a signboard. In 1630, the Pope had even approved the publication of the Dialogo. Galileo's original title of the work had been "About Ebb and Flood", since he thought them to be the decisive proof of the Earth's movement. Pope Urban VIII insisted on the final "Dialogo" title, which the Pope had proposed himself. Actually, Galileo complained about this title change to the Swiss lawyer Elia Diodati in August 1631, since he believed that his original title would gain a wider distribution. The actual publication, however, still took some more years to achieve. This cannot be attributed to the tardiness of the Pope, but rather to the handling of the manuscript by the person who was assigned the task of promoting it. This man was the Dominican Niccolo Riccardi, Galileo's adversary. He retained the preface and conclusions for more than a year and released them only after pressure from the legate Franscesco Niccolini and his wife Caterina, under the condition that nothing would be changed at all. But despite every concession by Galileo to later censorship, his ideas and those endorsed by the ecclesiastical view remained incompatible until the end.

The Dialogo

Galileo dedicated his work to the Grand Duke of Tuscany, Ferdinand II. In his introductory letter, Galileo wrote about the great book of nature, which guided him in his endeavours, and his objective of explaining first and foremost the structure of the cosmos. He put Ptolemy and Copernicus on an equal footing. The discussions in his tract simply centered on the scientific results of these two great thinkers. He even went so far as to proclaim that the results of his own deliberations were in reality those of the Grand Duke himself, because of the encouragement he had received from him.

Before he entered into the thick of the subject, Galileo introduced the reader to the scope of the work. He started off oddly by praising a beneficial edict, published some years ago, which had tried to stem the increasing belief that the Earth was moving rather than being motionless. Since his aim had always been to defend the truth on the world stage, he confessed that his opinion had been consulted prior to the publication of the edict. Nevertheless, during the preparation of his point of view, he had been forced to consolidate all arguments for and against Copernicus and had finally decided to side with the latter. He softened his stance by stating that although the new world model may perhaps not be quite superior to the traditional model, in which the Earth did not move, the arguments of the Peripatetics—a school which defended the teachings of Aristotle at all costs—did not hold up in the face of his own mathematical proofs. Galileo then outlined the structure of his book. He wanted to deal with three important subjects:

- the laws of motion on Earth leading to non-conclusive results about the motion of the Earth
- celestial phenomena and finally
- the tides.

He introduced the three persons participating in the Dialogo as close acquaintances of the past, of whom Giovanfrancesco Sagredo and Filippo Salviati both had in the meantime succumbed to the plague, while the third, still alive, appears in the Dialogo under the pseudonym of Simplicio. Their respective roles are as follows:

- Sagredo played the neutral host; he had been a leading political figure of the Venetian Republic.
- Salviati defended the Copernican position; he was a descendant of a wealthy patrician family of Florence and had studied in Padua.
- Simplicio defended Aristotle; it is not quite clear whether he stands in for a well-known Peripatetic or whether Galileo referred to a famous commentator of Aristotle from the sixth century.

The overall text is formally divided into four day-long sessions.

The initial discussions centre on very basic foundations that permit a scientific discourse in the first place: the two different substances of nature, the heavenly and the elementary ones, the nature of numbers as such, the number of dimensions and the fact that any progress in science always has a preceding deficit as its trigger. Salviati tries to strip the mystical from numbers by applying a deductive approach. And by employing geometric methods, he proves definitively that there can never be more than three dimensions. At the time of that discussion, of course, no one had yet heard about the string theory of modern physics, which requires ten or more dimensions, depending on which version one is considering.

Before turning to the subject of motion, the question of the perfection of a body is debated in some length, during which debate the defender of Aristotle claims that science does not always need strict mathematical proof to explain nature. In fact, Simplicio continues throughout the debate to cite his ancient master as sufficient evidence against any observed facts brought forward by Salviati, being refuted in turn by his adversary. In effect, Galileo finally uses this exchange of arguments to settle his disputes with Aristotle himself once and for all.

Aristotle

In truth, the whole discussion centred around the basic concept of the scholastics of the division of the world into heaven and Earth, into heavenly bodies that are proclaimed to be perfect, unalterable and indestructible and Earthly ones, which are highly imperfect, alterable and destructible.

Throughout the Dialogo, Galileo attacks the positions of Aristotle through the voice of Salviati, constantly being either contradicted or prompted by Simplicio. The arguments about the tides will be dealt with further down in this chapter at its proper place. Here are the main issues at stake:

The perfection of the world is based on the three-dimensionality of its bodies: Galileo laments that no strict proof of this exists. The Peripatetics base this assumption on the nature of the perfect number "three". "Three" signifies allness. This idea had already been suggested by the Pythagoreans in their number theories long before Aristotle. The number theory of the Pythagoreans had been consolidated by

Nicomachos of Gerasa. It is important to remember that in ancient times mathematics comprised subjects like arithmetic, music, geometry and astronomy (just as Kepler later tried to re-consolidate them in his world harmony). Astronomy in the end delivered the cosmological importance of numbers for the creation of the world. Nicomachos wrote[11]:

> All things in the world, which have been organised workmanlike, appear distinguishable and well organised due to numbers used by providence and reason, which have brought the universe into existence. Because the pattern had been preset just like a plan, controlled by numbers, already in the mind of the creator god, discernible only via numbers ....

Continuing with this line of reasoning, there also exist only three types of motion: circular, straight and composed of the first two. Of these, only one real circular motion truly exists and there is therefore only one centre around which it takes place. Galileo holds that there must be thousands of centres and thus thousands of circular motions belonging to the fixed stars. With regard to straight motion, Galileo evokes the primordial chaos, which, he declares, made use of straight movements to organise itself.

Bodies are either elementary or composed; elementary bodies possess natural impetus, composed bodies composed impetus, which may come from random causes. This Galileo had refuted by his own experiments. The velocity of downwards motion depends on the material composition of objects on the one hand, and is, equally, proportional to their weights.

Galileo's main attack on the ancient master, however, is a methodological one. He finds that Aristotle's explanations were drawn mostly from the mental goals he had in mind and were tailored to suit at them. He presumed the existence of what he wanted to prove—that it existed, in the first place—and then proceeded in a kind of circular argument.

The whole foundation of Aristotle's construct of ideas is said to be based on his claim of the existence of a body or an object surpassing all others in perfection. Straight lines are imperfect since, if space is infinite, no endpoint for such a line can be found, or—should space be finite—they can be extended beyond any hypothetical end point. In the end, all else—lightness, heaviness, volatility, immortality—is based upon this first supposition.

The Aristotelian proofs brought forward by Simplicio are of the kind: "Aristotle wrote it and therefore he proved it and therefore it is true (ipse dixit)"—his standard answer to every argument. And therefore Aristarchus and Eratosthenes were wrong because Aristotle thought they were. The answers to all questions that it is possible to ask are already contained in Aristotle's books. He could not have erred. And without Aristotle, there would be no scientific enterprise at all.

Galileo in the person of Salviati fights against this with his world of the senses, with common sense and probabilities derived from observation—a fight against paper, as he puts it.

From the geometrical nature of the universe, the discussion of the three eminent persons then turns to the question of motion itself, i.e. acceleration. There is agreement that the state of maximum slowness is the state of rest. From this state, an accelerated body acquires its increasing speed by passing through an infinite number of stages of slowness until the end point of the movement is reached, and this occurs on the shortest possible path. Salviati then goes a step further by asserting that a straight movement may well be transformed into a circular one and applies this to the motion of Jupiter.

Time

During the discussion about the incremental stages of slowness, Galileo, through his proxy Salviati, anticipates infinitesimal calculus, invented by Newton and Leibniz about half a century later.

This is what Galileo (Salviati) says about the time required for acceleration:

> As far as I seem to have understood, your principal plea is directed against the notion that a body should pass through those infinite preceding stages of slowness ad this in the shortest possible time, until it reaches its intended velocity after that time interval. There I would like – before I proceed any further – to eliminate these concerns, which is not difficult. I only need to respond, that the body, although it passes those already mentioned stages, it does not pause at any of these. Since such a passage does not require more than just a single instant, and since any such time interval contains an infinite number of instances, we will provide for a sufficient number of instances and assign to those infinite numbers of different stages of slowness each its proper particular point in time whatever the smallness of the time interval in question.

This quite remarkable explanation corresponds to what one may call "Leibniz Time". Past, present and future: what is past does not exist—not any longer. It is gone. Nobody can stretch out his hand to reach into the past. What will happen in the future does not exist—not yet. Nobody can stretch out his hand and fetch something from the future. What remains is the present, and this is as thin as a Leibniz time interval: infinitively thin.

This concludes the first day of the discourse.

During the next stage of the discussion Salviati and Simplicio quarrel about the relativity of motion with of course the latter defending the Aristotelian/Ptolemaic position that the Earth remains at rest eternally. The discussion centers on two major examples:

- The composite motion of an object falling from a tower made up of the downwards movement to Earth and the circular movement, as the object is carried along by the revolution of the Earth itself.
- The relative speeds of cannonballs, one fired in the direction of the revolution of the Earth and another in the opposite direction, and the differences that would result.

Both examples evoke thought constructs and experiments that lead ultimately to the Theories of Special and General Relativity. Let us start with the latter example. It reminds us of the famous Michelson-Morley-Experiment to measure the speed of light, already mentioned in the preceding chapter. However, in that case it was not the ether which was at the centre of the argument, but the combined motion of the light source and that of the Earth—just like the proposed experiment in the Dialogo with cannon balls. As we know, the Michelson experiment failed in its objective, since the speed of light always remains the same, regardless of the direction of motion of its source. The so-called Galilean Transformation does not hold in this case, but it would have held in the case of the classical mechanics proposed in the Dialogo.

The first example above leads us back to the question of the centre of the universe, already dealt with earlier in this book: a falling object from a certain height: what is its true movement under modern cosmological considerations?

The Theories of Relativity assert that there is no absolute system of reference. To illustrate this, let us again assume that someone stands on the roof of a house and throws a ball downwards. In classical terms this would mean: free fall, straight line, shortest distance. The geodesic would be a straight distance. But it is not as straightforward as that, as we know from the experiments in Budapest and Princeton (Chap. 4). There are additional influences. For one, there is the Earth's rotation (just as proposed in the Dialogo). Then, the Earth also is moving around the sun (Dialogo as well) causing the ball to move along an additional arc. Then again, the sun does not stay in one place (this is beyond the Dialogo), but moves about within the Milky Way. A distant star observes yet another circular movement …. and so on with galactic clusters and the entire cosmos. The centre of the universe is certainly not to be found on the roof of that building. The accelerated ball makes a complex movement along a geodesic, which—although it takes the shortest possible distance between two points in space time—is not a straight line.

Salviati tries to convince Simplicio of the first part of this reasoning by geometrical means, during which he illustrates that the composite movement of a falling body is made up of its downwards motion together with the circular motion of the tower from which the object is dropped. Simplicio takes a final stand on this subject, maintaining that we do not notice all this complexity in our everyday lives. Salviati refutes this argument by declaring that we are part of the system and would therefore notice only those components of motion that are external to our participation.

The discussion continues with other examples of cannon balls being shot upwards vertically and culminates in a theoretical calculation of the time it might take for an object falling from the moon downwards to Earth. Salviati applies to this case the classical law of falling bodies as we still know it today. To sum up, Galileo showed that objects on Earth moved as though their common movement with the Earth did not exist. In this way the second day was concluded.

On the third day, the discussion finally approaches the crucial subject at stake: the structure of the universe itself. Applying all of the mathematics and physics of the preceding deliberations as well as objective observations of the planetary movements, such as the Venusian phases, Salviati induces Simplicio step by step to develop a planetary system that puts the sun into the centre. On a drawing on paper, Salviati lets Simplicio fix the position of the Earth and then at another point that of the sun. Thereafter Simplicio is led to deduce the positions of all the other planets on the basis of actual celestial observations—until in the end the result becomes a chart of the heliocentric system proposed by Copernicus. Even though Simplicio is convinced by the pure mathematics of this model, he still refuses to accept its reality. Several questions arise:

Why, if it is so, had this model, which—as the moderator Sagrado interposes—is an age old one, first developed by the Pythagoreans, not convinced the majority of scientist in the past? Not to mention the ordinary public over all these years? Why were there so few followers?

It did not take long for Salviati to refute this argument: he cited his personal experience with audiences listening to his own lectures on the subject. He simply brushes away Sagredo's question, insisting polemically on the simple and unenlightened minds of the majority of people, while admiring those who had the courage to argue against the mainstream—like he himself.

Other intuitive arguments against Copernicus include the false conclusion that by the rotation of the Earth people would be carried from one place to the next, having lunch in Persia and dinner in Japan, which again Salviati brushes away as silly. A more serious matter is the problem that adversaries of the heliocentric model had with the weight of the Earth. To them, it is simply unimaginable that such a heavy body should be capable of rising and falling during its orbit of the sun. And a third objection is constructed with the purported fact that a person at the bottom of a deep well would never be able to follow the movement of the stars under which the Earth was passing because of the angular velocity of the rotation of the Earth. Salviati refuted both arguments by proving that the same objections were equally applicable to a geocentric constellation.

But what about the mountains, if the Earth was spinning around? Would they not suddenly become horizontal so that people would be able to walk to their tops, as though they were walking on a horizontal surface? Salviati cited the obvious fact that the Antipodes live upside down even on an Earth that is perfectly at rest.

The third day ends with the discussion of details that had escaped Copernicus in his publication, but these Salviati attributed to the lack of scientific instruments at his disposal at the time.

The fourth and last day of the convention: Salviati puts forward his strongest argument in favour of Copernicus but fails with respect to the standards of our time: the explanation of the tides. Galileo reached the limits of his theory of motion in trying to prove the agreement of his theory of motion with his celestial observations. He advances two important arguments:

1. If the Earth were motionless, there could never be ebb and flood.
2. He rejected the argument, put forward by some other researchers, that the moon and the sun were the cause of the tides.

Galileo tried to explain this natural phenomenon purely as a combination of the Earth's rotation and its circulation around the sun. According to him, the differences of the tidal ranges around the globe were caused by various surface features like the flatness of a basin or the depth of the ocean or crevasses etc. He compared the motion of the sea to the motion of water in a container carried by a ship. The water would rise and fall against the edges of the container as a function of its acceleration or deceleration, just as would happen with the oceans of the Earth. He even rejected the periodicity of six hours with regard to tidal changes as not being of natural origin, but only representing a statistical value that had been observed more often than other time intervals.

The session was concluded after four days. Salviati had dominated it with his ideas confirming the superiority of the Copernican world model with regard to the causes of motion, the constellation of the planets and the causes for ebb and flood. However, at the very end of the Dialogo Galileo, in his well-known way tried to again leave a loophole open: although all rational arguments were pointing in the direction of Copernicus, the whole discussion was no more than an attempt to compare different points of view, and if someone would turn up with serious arguments to refute the conclusions, then this person would be very welcome. He, Salviati (Galileo), had no intention to hurt the other side, and finally—since no one could ever penetrate the work of God's hands—it may well be—as would be the case with the tides—that they were caused by Divine intervention.

At various points in the Dialogo, mention is made of some mysterious "academic friend"—a person apparently possessing superior scientific insights. This person is never identified by name, and it can be assumed that this was a self-reference by the author, Galileo. It is improbable that this person was a certain Christopher Wursteisen from Germany, who had come as a student to Padua, and of whom it was said that he had introduced Galileo to the teachings of Copernicus.

Impact

After its publication and before its prohibition, the Dialogo found widespread acclaim in the more enlightened circles of science. On June 19th in 1632, his old friend and follower Benedetto Castelli, who had followed his teacher in the chair of mathematics in Pisa, wrote an admiring letter to Galileo, in which he reported on intensive discussions with other researchers about the book, among them Evangelista Torricelli, the later inventor of the barometer, but also about a negative reaction from Father Scheiner, who wanted to write a reply to the Dialogo. Castelli asked Galileo for several more copies of his book for wider dissemination. He also forwarded requests from other people, among them Giovanni Battista Ciampoli, the co-founder of the Sapiencia, the old Roman University, secretary at the Vatican and at that time still a person of trust to Pope Urban VIII. Castelli expressed his conviction of the correctness of Galileo's reasoning on the tides.

Other acclaim came from Fulgenzio Micanzio, a member of the Order of the Servants of Mary and a participant in the "Ridotto Morosini", an intellectual circle in Venice, to which Galileo also belonged. Micanzio's praise centred especially on the explanation of falling bodies and other motions given in the Dialogo. In his letter to Galileo from July 3rd 1632 he also exulted about the assault on Aristotle's position and that of the Peripatetics. In a further letter, dated September 18th of the same year, Micanzio deplored the developments which had taken place in the meantime. At that time, it had become evident that Galileo had become an object of persecution and the Dialogo was at risk of being forbidden. Micanzio, however, assured Galileo that he was convinced that his teachings would be admired and disseminated everywhere else in the world other than Italy.

In view of the clerical complot in the making against the Dialogo, Campanella warned Galileo that certain circles in the Vatican were drawing together a group of scholarly people to prepare a trial against him. Campanella, however, seemed confident that the Pope would be enlightened enough to resist these attempts. He even proposed that Galileo write to the Archduke and request that he, Campanella, should join in the defence of Galileo. This was from Rome on August 21st 1632.

The discussion heated up, and on September 11th Torricelli himself took up the pen to communicate the opinions of the scholars with whom he was conversing. Obviously Fathers Grienberger and Scheiner rejected the conclusions of the Dialogo, whereas Torricelli himself begged to be accepted into the Galilean "Sect" of followers, as he put it. Campanella's attempts to assist Galileo at higher levels had failed so far, as he had not been allowed to bring forward his arguments in higher places.

After Galileo had received the summons to appear before the Inquisition in Rome, he wrote a letter on October 13th 1632 to Cardinal Francesco Barberini, nephew and confidant of Pope Urban VIII at the Vatican, in which he expressed his astonishment about the demand and pleaded that Barberini use his influence with the Pope to amend things. He had been well aware of the envy and hate provoked by his writings in his competitors in science, but he had never expected a measure of this kind. Galileo lamented about his age and ill health and proposed two alternatives for proceeding: firstly, he was prepared to consolidate all his past and recent writings and explain his position, which he deemed in accordance with his faith, as a general defence. If this would not be acceptable, he was ready to face the Inquisitors of the town of his residence, Florence.

At the beginning of 1633 Galileo addressed a further letter to his correspondent Elia Diodati in Paris, in which he foresaw the prohibition of the Dialogo, of which already a thousand copies had been sold, and—as a possible result of the upcoming Inquisition process—his future incapacity to round off his life-long research into motion.

Inquisition

What followed next was one of the most-commented and most-documented court cases in human history; the subject of so many deliberations; its history of impact on the humanities, the evolution of the enlightenment, its abuse in the fight against the church in general and cherry picking of individual arguments in any discussion about the superiority of the human mind and natural science in particular so unique, that it almost seems superfluous to describe the details of the proceedings as such.

But still, for the sake of completing the figure of our hero and his public stand, a brief recollection of the sequence of events is necessary.

To sum it up:

The print editions of the Dialogo were confiscated and a trial was convened. Galileo's public renunciation of Copernicus' teachings took place before the highest Court of Inquisition on June 22nd 1633. The official justification for the trial was Galileo's demand to review ecclesiastical dogma, thereby attacking the church itself. For this reason, the ecclesiastical authorities were forced to make of him an example.

"Inquisition" denotes a legal procedure as well as the institutions that implemented the fight against heretics, in both the late Middle Ages and in early modern times. The Inquisition operated the entire time from its inception at the beginning of the thirteenth century until its practical disappearance at the end of the eighteenth century. For its use, a new form of court procedure was established in the late Middle Ages: the Inquisitorial procedure. Its objectives were primarily the conviction and conversion of heretics and less, for example, the prosecution of witches. During the Middle Ages, there was no superior office or permanent institution to carry out the task. The Inquisition simply acted then and there where the church deemed it necessary with respect to special local circumstances. This changed in early modern times. Initially, the Inquisition was conceived as an ecclesiastical practise, only within the church, but it was later adopted officially to serve as a model for secular trials as well.

The Inquisition tried to practise a rational type of evidence-based procedure, relying on witnesses and documentary proof. Its procedures were documented in protocols. One of its problems was that the prosecution served as judge at the same time. For an Inquisition trial to take place the following pre-conditions had to be fulfilled:

- A heretic would have to be identified.
- A relevant church representative had to become active.
- The support of secular powers had to be enlisted.
- An Inquisitor had to be named.

The procedure as such was based on the following pattern:

The heretic in question was admonished prior to any further proceedings to abstain from his false conviction.

If the accused failed to comply, then a date was fixed for the proceedings to commence, first witnesses were summoned and suspicious documents produced.

Thereafter, the accused was questioned and then judged. He could either renounce his evil ways and get away with some minor punishment or remain stubborn and receive severe punishment—in the extreme being burnt at the stake. The use of torture was later formally admitted as a last resort, but was used with restraint, depending on the personality of the Inquisitor.

The appearance in public of the Inquisition started to change at the onset of the modern area. Three different regional areas of responsibility were identified: the Spanish, the Portuguese and the Roman Inquisitions. Of relevance here is the Roman type. It was initiated by Pope Paul III in 1542 and called "Sacra Congregatio Romanae et universalis Inquisitionis". It was manned by a council of six Cardinals who were at the same time General Inquisitors. Its main task was to prevent the spread of Protestantism and associated propaganda in written form in Italy. One of its main instruments was the Index Librorum Prohibitorum.

The Index

The Index Librorum Prohibitorum (Index of Forbidden Books), also called Index Romanus, was a register kept by the Inquisition. Catholics who read these books committed a grave sin. Reading some of these books could lead to excommunication. The Index was published for the first time in 1559. Its final edition appeared in 1948 with appendices until 1962, covering in the end more than 6000 books. It was abandoned by the 2nd Vatican Council in 1966.

It was under the auspices of the Roman Inquisition that the Index was established. The indexing procedure itself followed the following pattern:

- First there had to be a complaint either from inside the Inquisition or from somewhere else about a publication worthy of being put on the Index.
- On the basis of two independent expert opinions, a decision would be taken to proceed or not.
- In the affirmative, a panel of consultants would analyse the expert reviews and draft a recommendation to the Cardinal members of the Inquisition. The Cardinals would then decide about the danger posed by the publication and would pass their judgement on to the Pope for final decision.

The result could be:

- Indexing
- Non-indexing without publicly acknowledging that indexing procedures had taken place
- Request for further expert opinion.

The Index itself was divided into three classes:

- The names of heretic authors
- Heretic works
- Forbidden writings of unknown authorship.

Galileo

Taking the above mentioned sequence as a guideline, things developed as follows:

Galileo was summoned for a first interrogation on April 12th 1633 in Rome. Two persons were present: the General Commissioner Father Vincenzo Maculano from Fiorenzola and the Prosecutor of the Holy Office, Carlo Sinceri. After the usual preliminaries concerning Galileo's personal data, they asked him to confirm the authorship of the Dialogo book, then recalled his visit to Rome in 1616 and his first encounter with the Inquisition and the reprimand by Cardinal Bellarmino (s. Chap. 6). Thereafter a lengthy questioning took place about his promise to abstain from publicly defending the Copernican position. Galileo at that time had pledged to obey, and as a result of this, Bellarmino had later confirmed in a letter to Galileo that he had not been condemned by the Inquisition after all. Consequently—and after his visit to Pope Urban VIII—Galileo had from that day on in essence refrained from any public discussion of the Copernican world model.

The interrogation then returned to the Dialogo book, and his questioners wanted to know whether he had obtained any permission to have it printed beforehand. Galileo explained that he had obtained permission to print in accordance with the regular process, citing the names of the authorized Fathers whom he had contacted and who had cleared it. He also insisted that the book did not contain a defence of Copernicus but rather a disproof of his theories. At the same time he confirmed, however, that he never mentioned his first reprimand in 1616 to the Masters of the Holy Palace.

After this, Galileo was released to his prison quarters in the Palace of the Holy Office and was asked to declare upon oath that he would remain silent about the proceedings.

The second interrogation took place in the congregational assembly hall on April 30th 1633 at the request of Galileo himself (according to the official protocol). After having been asked to make his statement, Galileo proceeded to claim that it had been a long time since he had last read his own Dialogo, that he had had time during the past days to attend to it once again, and that it now appeared to him that the contents of the book looked quite different from what he had remembered during his first interrogation on April 12th. He admitted that certain passages in it might indeed be interpreted as a defence of Copernicus with special emphasis on the discussions of the sun spots and the tides. He conceded that an external reader might have deduced from these passages the correctness of the Copernican world system, but that it had never been his original intention to prove this. He even had the presumption to claim that, were he to write the work again, he would have no reservation about completely dismantling the whole argumentation in favour of the traditional view of the world. He admitted that he had written the tract out of vanity in the first place.

After a short recess, Galileo returned with the suggestion that a supplement to the Dialogo should be written as a quite logical continuation, since at the end of the present edition, the three participants in the discussion had agreed to meet again to further contemplate the consequences of their previous elaborations. In this supplement, he could easily disprove the wrong impression given by the Dialogo up to this point.

The same day it was decided that because of the old age and the ill health of the defendant the palace of the Grand Duke of Tuscany would be designated as the further prison of Galileo, agreed upon by the gaoler of the Holy Office, Francesco Ballestra.

The third interrogation took place in the presence of the same witnesses as the second and in the same place on May 10th 1633. In this brief encounter, the Commissioner granted Galileo a delay of eight days to draft a written defence to the accusations, which the defendant agreed to produce, albeit not as an excuse but rather as an explanation of his original intentions. To support this stance, he produced an earlier testimonial by Cardinal Bellarmino with regard to his person.

Galileo drafted his defence on the same day. It contained basically three elements:

- His justification for not informing the Master about Bellarmino's letter of sixteen years before when he asked permission for the publication of the Dialogo; he had assumed that the Master would have been informed about the contents of this letter by the records of the Inquisition in any case, as these records were nearly identical to Bellarmino's letter to him.
- Furthermore that he did not intentionally and by deceit insert passages defying his earlier instructions but only because of vanity, and that he was prepared to correct these errors.
- And lastly he begged his judges to consider his ill health and age of 70 and asked them to forgive him.

In the meantime Francesco Niccolini had intervened with the Pope to ask for leniency with regard to Galileo, and the Pope had assured him that everything would be done for the relative comfort of the defendant with respect to his actual accommodation, but also regarding his future "imprisonment", possibly in a monastery for some time, still to be decided upon by the tribunal. Still, the Pope would leave any benevolence to the discretion of the Grand Duke to avoid interfering openly and setting any precedent or example. Niccolini reported the Pope's position to Andrea Cioli, secretary to the Grand Duke, in a letter of June 19th 1633. At the same time, he informed Cioli that the tribunal had already taken its decision to banish the Dialogo.

The final interrogation took place on June 21st 1633. Galileo made a statement in which he explained that he had been indecisive for a long time about which of the two world models, the Ptolemaic or the Copernican, was the correct one, but that after his convocation by the Inquisition, he had finally made up his mind that the Ptolemaic one was without any doubt true. He defended the writing of the Dialogo as an exercise to weigh the arguments for and against one or the other system in a systematic way. Since all of the scientific arguments favoured neither the one nor the other, he came to the conclusion that only superior teachings could give the final answer.

His final statement on the subject itself was that he no longer supported the teachings of Copernicus and that indeed he had never affirmed them since he had been instructed to refrain from this in 1616. Galileo signed his renunciation and was lead back to his quarters.

One day later the verdict was passed. In a nutshell it contained the following elements:

- Reference to the affair of 1616 and Galileo's solemn promise to never again defend the Copernican system
- Ascertainment that the Earth is the unmovable centre of the world and that the sun is not at the centre, and that the contrary is incompatible with the faith

- Investigation of the Dialogo book, concluding that Galileo had tried deceitfully to promote once again the Copernican system, while pretending to have delivered a semi-objective comparison of the two competing world models
- Galileo's confession that he had worked on the book for the last twelve years with the aim to disseminate the Copernican system out of personal vanity
- Reference to his written defence of May 10th.

The verdict found Galileo guilty of heresy and thus liable for all punishments ordained for such a crime. The Dialogo would be forbidden. However, if Galileo was prepared to renounce his error once again according to the form prepared by the council for this purpose, they would be content to replace the proper punishment with a penance comprising the recitation of the seven penance Psalms every week for the next three years. At the same time he was condemned to further incarceration as long as the council sees fit.

The verdict was signed by seven of the ten judges. Among the three who did not sign was Francesco Barberini.

Then Galileo signed the form, in which he confirmed the structure of the accepted world system in accordance with the scriptures, accepted his past position as being erroneous, promised to refrain from repeating such errors ever again and promised to follow all instructions regarding the penance laid upon him, on June 22nd 1633.

"I, Galileo Galilei, have renounced as preceded with my own hand."

Prior to the judgement Galileo's adversaries had already succeeded in removing the Dialogo from the market, according to a complaint in a letter of April 1633 from the scholar Gabriel Naudé to the philosopher Pierre Gassendi in France. In this letter Naudé, complained that the book was no longer available in Italy due to the machinations of Jesuit Fathers, notably Father Scheiner. This was the first attempt to identify Galileo as a martyr of science.

According to Viviani, who confirmed in his biography of Galileo that his master, as a faithful Catholic, had accepted his error with regard to Copernicus, Galileo was released after five months in confinement. However, he was not able to return to Florence because of the rampant plague there. So he had to content himself with exile in the dwelling of his good friend Archbishop Piccolomini in Siena, from where he finally relocated to the countryside near Florence.

This account is partially supported by another letter from Niccolini to Cioli of June 26th, in which Niccolini reported that soon after the verdict, the Pope had intervened to house Galileo temporarily in the Gardens of Trinita dei Monti in Siena. Niccolini meanwhile had started negotiations with Barberini to designate Galileo's own country house near Florence as the site of his exile, once the danger of infection with the plague had subsided.

Chapter 9
Final Years (1633–1642)

Energy or its Difficulty in an Enlightened Age

> In a closed system the total energy content as the sum of mechanical, heat or any other type of energy remains constant.

This is the first main theorem of the principle of the conservation of energy. It was first proposed—albeit in a slightly different form—by the German physician Robert Mayer.

Robert Julius Mayer was born on November 25th in 1814 in Heilbronn and died in the same place on March 20th in 1878. He served as the surgeon on board the Dutch freighter "Java" and left for Batavia on February 18th 1840. During the passage, he had ample idle time to reflect on the world and its causes. Among other contemplations, he thought about the warming of sea water through the movements of the waves. This was one stimulus.

The other came from investigations of the sailors' blood. The colour of the blood changed during the day. At the end of work at night, the blood contained less oxygen than in the morning. It was darker. Oxygen had been burnt. Heat had turned into work.

During the journey back he had another one hundred and twenty days to ruminate. As a result he realized that light, heat, gravity, motion, magnetism and electricity are all manifestations of one and the same elementary force.

When he returned home, he published his findings in a first manuscript in Poggendorf's "Annalen der exakten Wissenschaft" (Annals of the Exact Sciences). It was six pages of speculation. He even calculated the mechanical heat equivalent (the amount of heat necessary to heat 1000 g of water from 0 to 1° C). When he presented his theory to the eminent Professors Noerrenberg and Jolly in Heidelberg, he met with rejection. This led him to publish a revised version in Liebig's "Annalen der Chemie und Pharmazie" (Annals of Chemistry and Pharmacy) in 1842. This paper remained

© Springer International Publishing AG, part of Springer Nature 2018
W. W. Osterhage, *Galileo Galilei*, Springer Biographies,
https://doi.org/10.1007/978-3-319-91779-5_9

without response from the scientific establishment for the next seven years. Further revisions were rejected by all of the publishing companies that he contacted.

Mayer then resorted to self-publishing at his own expense. He disseminated his tracts to the most important European Academies without success. In England, Joule confirmed Mayer's theses by experiment but failed to mention Mayer's name as a reference in his associated publication. On the basis of Joule's findings Helmholtz produced a paper "Über die Erhaltung der Kraft" (About the Conservation of Force), again with no mention of Mayer. This triggered dispute about who had made the discovery of the Law of Energy Conservation.

Friends, acquaintances and his wife abandoned him. He was admitted to psychiatric institutions and tried to commit suicide. Later, he took up his medical practise again. After many, many years, the Royal Society finally announced that it had been Robert Mayer who had discovered the Law of Energy Conservation.

This law forms the basis of what is called thermodynamics. But energy plays an important role in many physical phenomena, in classical as well as in modern physics. Work is the product of force and distance, with its units the Newton Meter or the Joule (sic!). Potential energy with respect to free fall is the product of mass and gravitational acceleration by height, and is converted into kinetic energy when a body is dropped and is equal to half the product of mass and the square of velocity. These aspects were discussed at length in Galileo's "Discorsi," discussed later in this chapter—although not in these precise terms.

Seclusion

After a personal order was issued by the Pope regarding Galileo's strict isolation from the outside world, Galileo was allowed to return to his country residence "Gioella" in Arcetri near Florence at the age of seventy in 1633. He remained under house arrest and was prohibited from any teaching activities. He was also not allowed to go to Florence to visit his doctors for his painful hernia. But he was permitted to visit his daughters in the San Matteo monastery. The prayers of the seven penance Psalms were fulfilled by his daughter Suor Celeste, as long as she lived. Other social contacts were restricted. However, he could continue his less controversial research, but any publication was forbidden.

In a letter to Galileo of January 3rd 1634, the philosopher and Plato expert Girolamo Bardi, living in Pisa at that time, confirmed Galileo's relocation from Siena to his domicile near Florence. In the same letter, he congratulated and encouraged Galileo in his intention to resume publishing and informed him of his own intention to release his first lecture on Plato in the form of an address against Aristotle.

In March 1634, Galileo recounted in a short description his time in Rome and what happened to him in the brief time since his relocation to his country house to Diodati. Although he had a quite comfortable accommodation in the house of the ambassador of Tuscany, he still uses the word "dungeon" for this. He also indicated that he had already composed a new tract about mechanics (the Discorsi) while still in Rome. With respect to his trials on account of the Dialogo, he shrugged off the insults to his life and reputation. Four months later, he wrote again to Diodati, this time giving a rather long account of his sufferings. He complained about the fact that he had to spend his time in the countryside rather than moving to Florence, where he had hoped to join circles of friends and eminent people to discuss affairs of interest in science and the world. Indeed, he had had a visit from an emissary of the Inquisition reprimanding him, and commanding him to abstain from any such requests in the future on pain of being transported back to Rome. Then he reported the death of his eldest daughter, whom he had visited regularly, together with her sister in the monastery nearby. His isolation was documented by the continuing wrath of his persecutors and the interception of his correspondence with personalities abroad.

Well known representatives of science at Pisa University and members of the Jesuit Order continued to publish papers directed against the theories worked out in the Dialogo. However, Galileo did not hesitate to entrust Diodati with the optical lenses requested by Gassendi, which he had received earlier.

A letter from the Inquisitor Muzzarelli to Cardinal Barberini of March 10th 1638 informs about the day-to-day supervision of Galileo's life and habits. The reason for this letter was a request to allow the convict a visit to a small church in the vicinity of his dwelling. In this letter, Muzzarelli explains that Galileo's son had been hired and paid one thousand Scudi per annum by the Grand Duke to watch over his sick and nearly blind father and to ensure that the restrictions concerning his contacts and movements decreed at the trial were being carefully observed.

In spite of all this, Galileo, however, entertained extensive exchanges of letters with friends and scholars in Italy and abroad.

He maintained contact to Bernegger (Fig. 9.1), whom he wrote on July 15th 1636 in response to a request for optical lenses and even a whole telescope. Galileo promised to do his best concerning the lenses, but advised against sending a complete telescope to Strasbourg in these uneasy times because of its size. In the same letter, he reported on a visit by Elsevier to his home and the latter's intention to publish the Discorsi.

Galileo went so far as to correspond with the King of Poland, Ladislaus IV, in Warsaw. This letter of July 1636 was accompanied by a parcel with three pairs of lenses for observations of the Earth and the sky. He explained to the monarch that he had been punished because of erroneous theories about the world system, which he condemned in this same letter as being even more corrupt than the writings of Luther and Calvin (probably to cover himself in case the correspondence got intercepted).

In August 1636, Galileo addressed the Dutch States-General to report his method to determine the exact degree of longitude of any location, which he wanted to make accessible to the general public. He proposed to have this method be investigated by eminent scientists of their choice, since he himself was unable to travel because for reasons of heath and because of the constraints imposed by the verdict of his trial. At the end of this letter, he asked for some sort of compensation for his efforts.

Galileo also received visitors, among them such prominent figures as Thomas Hobbes and John Milton, and later, from 1641 on, his pupil Benedetto Castelli.

Fig. 9.1 Peter Aubry:
Matthias Bernegger

He had long suffered from eye problems and became completely blind in 1638.
It is speculated that this was the result of his unprotected observation of the sun in
earlier years.

Discorsi

Despite all of the setbacks that he had suffered at the hands of the Inquisition,
Galileo had started work on his magnum opus in July 1633, while still confined in
Siena. He called it "Discorsi e Demonstrazioni Mathematiche intorno a due nuove
scienze" (Discourses and Mathematical Demonstrations about two new Branches of
Knowledge) or in short form "Discorsi". Once finished, he found out that publication

in the sphere of influence of the Catholic Church would be impossible. Thus, the rest of the world learned of the work only after Matthias Bernegger had translated it into Latin and published it under the title "Systema cosmicum" by Elsevier, printed in Strasburg by David Hautt.

It seems, however, that Galileo was not entirely happy with Berneggers boldness. This was expressed in a letter of January 19th, 1634 from Pierre Gassendi, the French scholar and philosopher, who conferred with both Galileo and Scheiner. In this letter, he announced the dispatch of a book by a Maarten von den Hove, a Dutch astronomer from Leyden, who taught the Copernican world model. Gassendi also requested high quality lenses for their telescope as these were obviously not available in his country, quite the contrary to the lenses to which Galileo had access.

Bernegger himself confirmed that he was revising and translating the Discorsi in a letter to Elia Diodati on February 24th 1634 and hoped to have it printed in the summer 1635 (which he did). The Frankfurt bookseller Clemens Schleich would pay for the costs and distribute it. Bernegger, however, mistook Galileo's move to Florence as an act of liberation.

There was also an Italian version in 1638, again by Elsevier in Leyden. It is interesting to note that Viviani alleged that Galileo was unhappy about the fact that so many versions of the Dialogo had been disseminated in different languages beyond the Alps and that he was in no position to suppress them, since he had renounced it in the face of the Roman censorship. One can only speculate that Viviani tried to keep his own hands clean by qualifying his master's opinion in this way.

The structure of the Discorsi and its literary method are modelled along the lines of the Dialogo: the discussion is subdivided into four day-long sessions, and as protagonists the same actors are introduced: Sagredo, Salviati and Simplicio. The major difference from the Dialogo is that the subject matter of the Discorso does not deal with competing world views and thus would not have incited new controversies, but the whole book—although conceived as a form of discussion between scholars—really claims to be a kind of textbook, just as the Book of the Elements by Euclid [11] had been in its time. Just as Euclid structured his theorems into a succession of "construct" and "proof", Galileo structured a later part of the Discorsi as a succession of "theorems" and "propositions". Furthermore, the text is also almost completely free of any polemic against Aristotle.

Galileo by now was a scientist of international renown, and he was well aware of that. He articulates his ideas through the words of Salviati with the occasional reference, once again, to the ominous "academician, our friend". In sum, these are the contents of the book:

The first day is consumed by a detailed resumption (with respect to De Motu Antiquiora and Dialogo) of the problem of the free fall.

The second day deals with statics, especially with regard to the inherent stability of a body with respect to its own weight.

On the third day the subject of motion, especially acceleration, is taken up again, and here we find the connection to Robert Mayer and the transformation from potential to kinetic energy.

The book concludes on the fourth day with mathematical descriptions of thrown objects.

The discussions starts with the finding that at present more and more machines and mechanical devices are being constructed and released for common usage, and the question arises about the causes that propel these apparatus or even natural phenomena. The disputants agree that the basis for all this is to be found in the laws of geometry, particularly in those dealing with proportions. Surprisingly, Salviati takes up the matter of real machines being imperfect with regard to ideal machines, thus anticipating a debate that many years later ended in the formulation of the Second Law of Thermodynamics, formulated by Rudolf Clausius in 1865. This law can be expressed as:

All natural processes are irreversible.

This leads to the speculation that nature has obviously provided for some limitations with respect to the composition of its objects. These limits manifest themselves in the size and weight of objects, among other ways. These limits are equally applicable to artificial man-made objects.

After these general introductory discourses, the three continue to take up the matter of motion once again, this time starting with the relationship between vacuum and motion, until they arrive once more at the motion of cannon balls. Other aspects that are covered deal with the different resistances that surrounding media offer to free falling objects within them and their specific weights, finally arriving at Archimedes' principle without mentioning it as such. A point of contention arises between Salviati and Simplicio regarding the weight of air itself. Salviati finally proposes two different experimental methods to measure the weight of air and thus resolves the conflict.

From free fall experiments, the discussion then turns to movement down an inclined ramp, the movement of a pendulum and the laws governing the latter. Other aspects relating to this complex subject matter of movement include friction and acoustics with regard to the vibration of strings on musical instruments. This concludes the first day.

The second day commences with skull and bones. It is about the limits on the possible weight of hypothetical giant animals and the explanation as to why huge fish do not collapse in water due to their own weight. They then proceed to apply their reasoning to man-made objects, such as pillars and bridges, and thereby arriving at the basic laws of statics and leverage.

After the rather brief discussions on the second day the third day starts out with a subject already abundantly discussed in previous papers by Galileo: natural motion and acceleration. They again arrive at the incremental accretion of speed, thus touching again on the principle concepts of infinitesimal calculus still to be developed years later by Leibniz and Newton. By again delving into the example of perpendicularly thrown objects, notions of potential and kinetic energy come to the surface of the discourse without being formalised explicitly. But in the end, the whole train of reasoning comes close to the foundations of the First Law of Thermodynamics. However, when it comes to the question of the real cause of acceleration in free fall, Salviati has to give up.

It is on this third day of the discourse that Galileo introduced his concept of "Theorem—Proposition". He has this concept introduced by Salviati in several examples that present the results of the research by the ominous author already mentioned earlier (Galileo himself):

Theorem 1: The time required to cover any distance by an accelerated body from rest up to a uniform accelerated movement is equal to the time that would be needed by the same body to cover the same distance at a mean uniform velocity that is half that of the maximal speed reached during the previous acceleration (Proposition: geometrical proof).

Theorem 2: If a body falls down in a uniform acceleration from rest, then the distances covered in certain time intervals are relative to the square of the time intervals (Proposition: geometrical proof).

The fourth day is dedicated to the physics of the thrown object.

Theorem 1: A body subject to a uniform horizontal and at the same time uniformly accelerated movement describes a half-parabola.

The Discorsi had been completed already in 1635. In a letter to Diodati of June 9th, Galileo informed him that he had given a copy of the first two of these dialogues to Prince Mattia (de Medici).

He took up the matter in a letter to Micanzio of June 21st 1636, one year later, confirming his dealings with Elsevier, who apparently had been staying at that time in Venice, in connection with the optical lenses for Bernegger and the publication of the Discorsi. At the same time he asked Micanzio to arrange that Elsevier might reprint two of his works, which were in steady demand in Latin; works dealing with sun spots and floating bodies in water, obviously extracts from his other major publications. He also wanted Elsevier to issue a copy of the directions for use of his proportional compass again in Latin for Bernegger. This communication was followed up with a further letter on August 16th accompanied by his book about motion (De Motu). In both letters, Galileo referred to one of his nephews, Alberto Cesare, who lived in Bavaria, and whom he wanted to support financially and whom he invited to visit him and stay with him in Arcetri (Fig. 9.2).

Last Efforts

Viviani stayed with his master as his private secretary and assistant until late 1641, before being replaced by Evangelista Torricelli. Galileo's desperation over his blindness becomes evident in a letter he dictated to his assistant and addressed to Diodati on June 6th 1637, in which he complained about his inability to read or write, to continue observations of Jupiter or to calculate and revise the associated mathematical tables. He had promised these efforts together with a proper description of the exact determination of the degree of longitude to the Dutch mathematician Martin

Fig. 9.2   La Villa Il Gioiello, Galileo's Last Home in Arcetri, today (© Cyberuly)

Hortensius. But Galileo still did not despair completely. In another letter to Diodati of November 7th of the same year, he asked his correspondent to "stretch out his helping hand" to relieve him of his sorrows so that he would be free to again defend himself against his adversaries on the basis of all his scattered physical and mathematical papers.

Two days earlier, on November 5th he had written to Micanzio with similar complaints but still quite confident that he was actually compiling a register of his astronomical observations past and present.

Viviani recounted the following episode related to Galileo's blindness:
Because he could no longer supervise the publication of his latest astronomical observations, he handed over the material to another pupil of his, P. D. Vincentino Renieri, who later became mathematician in Pisa, to revise his tables. Renieri published some tables about the positions of Jupiter's satellites in 1639. But most of the letters and records he inherited disappeared after his death and were presumably stolen.

According to Viviani, Galileo endowed the Discorsi together with copies of his other writing about motion (Motu) to the Count of Noailles, who passed by in 1636 on his journey back from Rome, and who had them printed in Leyden. Viviani then continued to narrate the origination of the Discorsi and the role he and his successor Torricelli had played when their master had dictated his thoughts to them. According to Viviani, Galileo had still other projects in mind, but could not finish them before his death. Among them were additional theorems and propositions, refutations of sentences and opinions of Aristotle, and ideas about percussion. Galileo was busy with the theories of music, painting, sculpture and architecture.

Finally the question of the correct world model came up again in a letter to the Florentine delegate to Venice, Francesco Rinuccini. In the letter Galileo started right away by claiming that the Copernican system was utterly false and that there should be no doubt about its falsity, since theologians had proved this on the basis of the Holy Scriptures. On the other hand he did not abstain from stating at the same time that both Ptolemy's and Aristotle's speculations were equally insufficient as proofs for their own world models. And, in an aside, he indicated that any further argumentation pro or against could be found in his unfortunate Dialogo.

Galileo died on January 8th 1642 in Arcetri, before completing any other new work after the Discorsi.

The solemn burial in a pompous tomb envisaged by the Grand Duke was prevented. The Grand Duke had previously visited Galileo on several occasions on his sick bed. In a short note from Giorgio Bolognetti, the Apostolic Nuncio to Florence, to Barberini on January 12th 1642, Bolognetti had informed his correspondent about the intention of the Grand Duke to erect a monumental sepulchre for Galileo similar to that of Michelangelo, just opposite of that of the artist. In reaction, Barberini instructed Muzzarelli in a letter of January 25th to prevent this, since this would have hurt the feelings of any true Catholic. Any inscription on the tomb should carefully avoid damaging the reputation of the Inquisition tribunal.

Galileo was thusly buried anonymously in Santa Croce in Florence. It took thirty years before his tomb was marked by an inscription.

Chapter 10
Conclusions

Time and Space Revisited

The European Middle Ages are generally thought to encompass a time span dating from the fourth to the fifteenth century. The designation "Middle Ages" is somewhat of a misinterpretation of history. A few scholars had started to regard the preceding cultural era as something far inferior to their own. Fascinated as they were by Greek and Roman antiquity, the humanists viewed the whole European past in the light of these ancient civilisations, which were followed by a long period of barbarity, in which ignorance and hatred of the beautiful prevailed. In short, in their eyes the Middle Ages represented a "middle" epoch of decay situated between the two blossoming eras of antiquity and its re-birth in the Renaissance. The "middle age" was considered to be so mediocre that even its most important achievements were given only derogatory names, such as "gothic", a term coined from the name of the barbarian tribe of the Goths.

The Middle Ages differ principally from all other epochs in its indissoluble link between church and society. Society in the Middle Ages understood itself as "ecclesia". This term, which is translated only inadequately with "church", signifies a Christian community living wholly with a view towards eternity. Modern concepts of faith and religion are insufficient to describe such a society. Faith presumes a kind of choice, which, however was non-existent in the Middle Ages.

To describe medieval dynamics, scientists have divided the Middle Ages into smaller sections. They talk about the High or Late Middle Ages, the Early Middle Ages or the Early Renaissance. In the German and English contexts, the terminology is similar. The time span between the eleventh and the thirteenth century is called the High Middle Ages. In French speaking cultures, all boundaries in time shift by two centuries ahead. The time span between the fifth and the eighth century is not the Early but already the High Middle Ages. The period between the fourteenth and the fifteenth century is already called the Low Middle Ages.

Space and time, life and death, heaven and hell, cosmos and Earth—everything had been conceived of and measured by yardsticks radically different from those of

© Springer International Publishing AG, part of Springer Nature 2018   131
W. W. Osterhage, *Galileo Galilei*, Springer Biographies,
https://doi.org/10.1007/978-3-319-91779-5_10

antiquity. This was a slow process, and to measure space and time means at the same time to command them. The Earth and the celestial bodies were creations of God and in accordance with these, the whole of creation had been built on antagonisms: above and below, centre and periphery, spiritual and carnal. Above, the kingdom of God is situated, below, the realm of man und of transience. The centre is good, the periphery a place of uncertainty, if not of evil. The world map shows the Christian regions—basically Western Europe—at the centre. The fringes of the world are populated by heathens, monsters and legendary creatures.

Schools offered a formal setting for the transmission of knowledge. Medieval schools cannot be compared in scholastic practices to schools in Greek or Roman times. They were situated near cathedrals and in monasteries and served to educate clerics and some selected lay people. Education and the clergy mixed over time. The cathedral and monastery schools remained strongholds of culture wide into the twelfth century.

Clerical instruction used reason to fathom revelation and the marvels of creation—being well aware that the human mind was prevented from a true perception of God. This conception slowly changes from the twelfth century onwards with the advent of the universities. Originally the designation "Universitas" meant simple a corporation. Brotherhoods, communes and guilds were all "Universities", whereas in common language only one "Universitas" is meant: "the University of Masters and Students". This model was developed during the twelfth century in Bologna, Oxford and Paris and later spread all over Italy, England, Spain and Portugal, later during the fourteenth century throughout the German Empire and Hungary. Kings and Princes played a significant role in the foundation of universities. For them, these became the centres of knowledge that served their own interests.

The academic degree of "Baccalaureus" opened up the door to higher faculties such as theology, law and medicine. A doctor's degree crowned the exploration of a special field of science. Universities were closed corporations and guarded their privileges. When members of the Franciscan and Dominican Orders joined them, they initially remained loyal to their orders, but soon widened the intellectual horizon of the universities by contributing non-Christian philosophers like Aristotle, Maimonides or Avicenna to the curriculum.

This was the world at the end of the Middle Ages, when Galileo stood before the threshold of the Modern Age.

René Descartes

Spiritual changes went hand in hand with political ones. Nikolaus von Kues (Cusanus), Erasmus and Calvin had prepared the ground just as Copernicus, Paracelsus and Kepler did. Calvinistic colleges spread unfiltered rationalism. Man as the bearer of reason took his place at the centre of the spiritual contest. Because of his rational aptitude, he found criteria for doubting received knowledge of natural laws and the tangible. The doubter himself became the yardstick for all things. Indepen-

Fig. 10.1 René Descartes

dently of revelation, he assumed the task of finding truths valid for the whole of mankind. The notion of faith was replaced by ideology.

Renatus Cartesius (1596–1659) (Fig. 10.1) is held to be the founder of the modern rationalism carried further by Spinoza, Malebranche and Leibniz. Cartesianism is another word for rational thinking. He coined the famous expression: "Cogito ergo sum" (I think and therefore I am) as the foundation of his metaphysics. He also developed his concept of the two substances that interact with one another: spirit and matter, today known as Cartesian Dualism as opposed to the dualistic natural philosophy of Newton, which teaches the interaction of active immaterial forces of nature with absolutely passive matter. Descartes also founded analytical geometry, linking algebra and geometry.

Descartes' name is forever connected to the coordinate system most in use world-wide: the Cartesian coordinate system, which is an orthogonal system. Descartes first made its use in two- or three-dimensional space popular. In two dimensions, the two directional axes are placed perpendicularly orthogonal to each other, thus meeting at an angle of 90°. The coordinate lines are straight lines at a constant distance from each other. The horizontal axis is called abscissa, the vertical one ordinate. The word ordinate is etymologically related to ordination in the churches, meaning the introduction of a person into a clerical office. It still lives on in the French word for computer: "Ordinateur", which indeed still carries a semi-religious meaning. The word was chosen by IBM France instead of the better translation "calculatrice" after consulting the teacher Jacques Perret, who suggested "ordinateur", a theological term meaning "The one who puts things in the right order". Thus theology still returned to mathematics in some manner during the twentieth century.

There is no evidence of direct communication between Galileo and Descartes, but in Descartes' correspondence one can find various allusions concerning the Galileo "case".

At the end of November 1633, Descartes wrote to Marin Mersennes in Paris. Mersennes (1588–1648) was a French theologian and mathematician who became acquainted with Descartes at the Collège Henri-IV. He became a member of the Order of St. Paul. While initially following scholastic teachings, he later changed sides to become a critic of Aristotle. In 1623 he visited Galileo and Descartes and thus became a facilitator of contacts between two of the most important scholars and scientists of his time. In this letter, Descartes writes about his own world system, which he still had not completed, but he had heard about Galileo's world system (Dialogo), which he tried to obtain in Leyden, since apparently all Italian copies had been burnt. He was well aware of Galileo's tribunal and its outcome. Since he himself, Descartes, stood in for the same model with a moving Earth, he had no intention of suppressing passages in his book pertaining to this aspect. He would rather refrain from publishing his book at all than to release a mutilated version of it. He then asked ironically for one more year of patience before he showed his tract to Mersennes, hoping in the meantime that his correspondent would abstain from sending a bailiff to confiscate his writings. In fact Descartes' works were later all put onto the Index in 1663.

In a further communication, Descartes wrote to Mersennes in February 1634 from Amsterdam referring again to his work, which he was at the brink of destroying altogether, in view of what had happened to Galileo, in order to prove his obedience to the church, but he requested Mersennes' opinion of the mood in France. He also referred to the role of the Jesuits in Galileo's condemnation and the book of Father Scheiner. In fact, after having read it, Descartes believes that a man like Scheiner himself must at heart be a follower of Copernicus, despite denying it.

Half a year later, Descartes finally held a copy of the Dialogo in his hands, loaned to him for some thirty hours by Isaac Beeckmann, a Dutch philosopher and friend of his. His criticism in a further letter to Mersennes in August 1634 centred basically on Galileo's explanation of the tides, but he also expressed some doubts about the exactitude of the free fall calculations and other theoretical motion experiments. In general he complained about the style of presentation as being to digressive at times.

On a trip to Endegeest on March 31st 1641 Descartes reported to Mersennes that he had been in contact with Antoine Arnauld, a scholar and teacher at the Sorbonne, critical of both the Jesuits and the Protestants and a representative of the Catholic reform movement, who had had some objections to his philosophical writings. Descartes was confident that his writings were not in contradiction to Catholic teachings and that he would not suffer any consequences as Galileo did. He claimed that the adversaries of Galileo had just blended Aristotle with the Bible and constructed their argumentation on this basis. He hoped that people at the Sorbonne would give a favourite verdict on his own writings.

Descartes' writings, including his most important work "Discours de la méthode pour bien conduire sa raison a chercher la verité dans les sciences ...." (Essay about the method for using one's reason well and how to look for the truth in science ....) and comprising cognitive science, ethics, metaphysics and physics, were all banned in 1663 after his death in 1650. Even during his lifetime, he had to flee his adopted country Holland for France and England for fear of persecution.

The rigorous rationalism of Descartes and Erasmus spread only slowly and with difficulty in the rest of Europe. In 1703 Isaac Newton was nominated president of the Royal Society. He remained in that office until his death in 1727. In his later years he lived withdrawn in a house in London, where he operated a small observatory. His favoured studies then comprised ancient history, theology and mysticism. John Maynard Keynes said that Newton was not "the first of the enlightenment", but "the last Magician". Keynes had acquired part of the written inheritance of Newton, originally handed down to Newton's niece Catherine Barton after Newton's death, in an auction at Sotheby's in 1936. The bulk of these manuscripts dealt with research in alchemy.

Galileo's Tangible Scientific Achievements

There were inventions and discoveries, there were stances and teachings, and there was posthumous glory acquired by controversy and suffering at the hands of the powers of his time. Leaving aside the latter details to another section later in this chapter, Galileo's tangible achievements have to be weighed against the technological and scientific resources available during his lifetime.

Galileo has been credited with these major inventions:

- Hydrostatic Balance
- Galileo's Pump
- Pendulum Clock
- The Sector
- Galileo's Thermometer
- Telescope.

His inventions not only demonstrated his genius as a scientist but also were proof of his engineering skills. Although he certainly engaged craftsmen to construct his devices, he certainly worked on them himself while realizing his instruments. In any case, he devised and supervised the technologies in question. This goes as well for improving such simple laboratory devices as the inclined ramp. In some cases, he put his engineering to use to earn some money on the side. On the other hand, although Galileo has been credited with some inventions, they were not entirely of his own making.

The hydrostatic balance was inspired by a tale of Archimedes' discovery of specific weight. And his pump, for which he was granted a Venetian patent, was based on the Archimedean screw. The invention of the pendulum clock is generally credited to Christiaan Huygens, who published its workings in 1657 after Galileo's death. However, Viviani reported that his master had been working on such a device already in 1641, but was unable because of his blindness to complete his studies. Unique is the invention of his sector or compaso. His development of a thermometer before the existence of absolute or relative temperature scales can be regarded as a major breakthrough, although his apparatus should better be called a thermoscope.

The history of the telescope is well known. Galileo was but one who improved earlier constructions but who certainly takes all the credit for its application to astronomy and the resulting discoveries of new celestial bodies. This takes us from engineering to science itself.

These are his main publications:

- Siderius Nuncius
- Saggiatore
- Dialogo
- Discorsi.

They have been discussed in detail in the previous chapters. There is no doubt that they proved to be the foundations of his fame during his lifetime long before the conflict with clerical authorities triggered the history of his glorification. Galileo was accepted as an eminent scientist in many fields, ranging from philosophy, physics, astronomy and others. At that time, scientist were of a universal type, capable of a world view encompassing nearly everything, something that in our present times of specialisation would be attributed to a so-called lateral thinker. However, he did not alone possess this kind of qualifications. Many of his contemporary colleagues were similarly endowed. On the other hand, there were those like Markus Welser, Lord Mayer of Augsburg, a friend and benefactor of Galileo, who begged in a letter from May 20th 1611 "to be spared for a while to follow Galileo along the path of the Earth's movement, because he had difficulties to shackle his mind in as much."

The really outstanding tangible achievements were his astronomical discoveries, first of all the discovery of Jupiter's satellites and his observations of the surface of the moon. If at that time something like a Nobel Prize for physics had existed, Galileo surely would have been a candidate for it because of this.

The discussion of Aristotle and motion as well as the competition of the two world systems is another matter. Here, Galileo did not present a unique position. Many other scholars of his time were heavily involved in these quarrels; none of them had proposed the heliocentric system himself, including Galileo. It was there to take it or leave it. But the time was ripe. Unfortunately, because of his already well established standing in the scientific world and in society, Galileo's word had quite a different weight than that of those in inferior positions. But his standing in society and connections to the highest clerical circles not only forced the latter to act as they did but also in the end protected him from a similar fate to that of Giordano Bruno.

However, there are other more fundamental aspects of Galileo's contribution to modern science. He also tried to popularise science by using everyday language, the "Volgare", instead of the baroque style of presentation in favour with his contemporary colleagues.

It is unfortunate that his drive for a science promoting the full application of reason, which radiated from his works, was superimposed by the never-ending discussion of morality, which today seems to have been the only source and foundation of his fame. The discrepancy between Galileo's approach and what was expected from a scholar of his time becomes obvious from one comment in one of the expert opinions drafted for his tribunal, which reads: "The author claims to have discussed a mathematical hypothesis, but at the same time he awards it with a physical aspect, a thing that a true mathematician would never do."

Of course, much of his thinking and fighting during his career centred on Aristotle and his followers. There seems to be no doubt that Galileo saw himself as an equal, a pillar of natural philosophy as great as Aristotle was regarded to be. He was at the same level as the ancient master. At first, he followed his Florentine teacher Ostilio Ricci and belonged to the scholars who rejected the literalism that prevailed in scholastic circles. To the critics belonged people like Johannes Buridanus, member of the Ockham circle of Nominalists, Nicolas d´Oresme, Bishop of Lisieux and Albert of Saxonia, Buridanus' pupil. They were called high scholastics. It was their Impetus Theory which first attracted the attention of Galileo before he started out to develop his own experimental mathematical method to study the measurable properties of motion.

A further source of stimulation came from Giovan Battista Benedetti, a successor of the Buridanus movement who had published a tract called "Demonstration of Properties of local movements against Aristotle and all Philosophers," which was critical of Aristotle's theory of motion. Benedetti had been a pupil of Niccolo Tartaglia, the mathematical master from Brescia who had been interested in free fall, the throw and trajectories of projectiles. All this meant that the young Galileo could resort both to writings of the ancients and at the same time to those of empirically oriented authors of his time.

In later years, his contributions did not exhaust themselves in opening up regions hitherto regarded as being unchangeable and thus inexplorable to the human eye from a mathematical and physical point of view. He went further in demanding that people should participate in his observations and reports. His sharp eye and deductive capabilities, trained by experiments with inclined planes and swinging pendulums, simple and complicated measuring tools, made it possible for him to also interpret observable changes in the skies. The result was the "Siderius Nuncius".

Galileo was convinced that his methods would allow for impartial exploration of nature, as he once confessed in a letter to Cesi. When it later came to crisis, he tried to defend his position against the assaults of clerical scholars by proposing a clear separation between the scope of the sublime (a phrase from Nikolaus von Kues), i.e. the Divine and revelation, and the scope of natural phenomena, where scientists should be left alone. Neither side should interfere in the scope of the other. In a communication to the mother of the Archduke Cristina di Lorena in 1615, he explained that theology indeed can claim to be of "highest authority", but only concerning its proper subject. It therefore should not try to adduce from this the claim that it is in possession of absolute truth at all. He asserted further on:

> Therefore its professors should not arrogate the authority to give orders in professions, which they had neither practised nor studied.

He compared the role of theology to that of an absolutistic Prince, who should have no interest to deal with the day-to-day business of medics and builders, since he would endanger people or buildings because of his lack of specialised knowledge.

His framework of thinking was founded on effective rational grounds. In the "Dialogo" he proclaimed the equality of Earthly and heavenly movements, deducing that the movement of the Earth constituted a completely normal occurrence—even before describing it mathematically. In the end, he managed the transition from speculative to verifiable science. The abstract concept of matter was replaced by the notion of its experimental objectivity as a system of physical forces. This approach and its presentation as a method of research in natural science remains a common theme throughout his work and has since reached the status of general validity.

Against Sarsi he had written:

> "Philosophy is written in this grand book—I mean the universe—which stands continually open to our gaze, but it cannot be understood unless one first learns to comprehend the language and interpret the characters in which it is written. It is written in the language of mathematics, and its characters are triangles, circles, and other geometrical figures, without which it is humanly impossible to understand a single word of it; without these, one is wandering around in a dark labyrinth."

This was written as early as the Saggiatore, and some people claim that the Saggiatore was the first manifest of natural science.

Historical Impact

Although Galileo's tangible scientific results were never forgotten and kept in good memory, the citation of his name invariably invokes mostly one thing: his conflict with the church and thus the evil the church did to science and mankind as a whole, discrediting itself at the same time as a reasonable institution.

During the nineteenth century, a number of biographies and essays about Galileo were produced each trying to position him either as a symbol of enlightenment or anti-clericalism under positive or negative moral signs. One famous theatrical piece is Brecht's "Galileo Galilei", depicting its chief character as both being scientifically

Fig. 10.2 Scene of Berthold Brecht's "The Life of Galileo Galilei" (Bundesarchiv, Bild 183-K1005-0020/Katscherowski (verehel. Stark)/CC-BY-SA 3.0)

obsessed and later manipulable by those in power (Fig. 10.2). But Brecht's drama was no more historically correct than Schiller's "Wallenstein's Camp".

This logic was prevalent throughout the development of the enlightenment and remains a standard reflex today.

However, there are other voices trying to provide a somewhat different interpretation.

Galileo was rehabilitated on November 2nd 1992 by Pope John Paul II. The records of his trial were made accessible to science in 2008. There were some interesting conclusions.

First of all, it seems that both were in error: Galileo on the scientific side, the curia regarding theology. The Inquisition did not realise that the contradiction between heliocentricity and the scriptures was only an apparent one. And Galileo was in no position to prove scientifically that his model or the Copernican world model was in fact correct. It had taken a long time for the Inquisition to act and demand Galileo's revocation of his conviction that the sun was the centre of the universe. But the crux of the argument was not what is obvious. What is obvious is that everyone can see every day that the sun indeed rises in the east and sets in the west. It does not stand still. Also obvious is that an Earth moving around the sun is simply not part of everyday experience either.

After Copernicus' publication, both world models, neither of which were proven scientifically according to today's standards, were for example taught in Spain in Salamanca in 1561. The Inquisition initially tried to nudge Galileo into being more careful with his apodictic statements. But in the end Galileo was too proud to adhere to this advice, and thus it came to the official trial. By the way: the verdict had never been signed properly.

In fact, the Copernican world model never had been a serious problem for church and Pope. Galileo tried to prove the motion of the Earth by resorting to the phenomenon of the tides. But the tides were a bad example, unfit for delivering the truth. This is standard knowledge today, and therefore on this point the Inquisition was right. Galileo even rejected the proposal to have his results categorised as a simple hypothesis until he had indeed to revoke them completely. This suggestion by Bellarmin is nowadays common practise: every scientific assertion must at first be regarded as a hypothesis. This notion was first formulated by Carl Popper in 1934 in his "Logic of Research". Galileo did not accept this idea then. To him, every thought emanating in his brain already presented some kind of truth, no matter how strange it might seem to others.

In 1908 the French physicist Duhem commented on the Galileo trial, maintaining that the scientific logic had been on the side of the Inquisition. He went so far as to state that, even if the hypotheses of Copernicus would have been able to explain all known phenomena, it could have been concluded that the hypotheses might be true, but not that they are coherent. To accomplish this, one would have to prove that there is no other system explaining the same phenomena as well or even better. The Theory of Relativity is one example.

The legend that the Inquisition had broken an old man, who at the end of his life would whisper on his death bed the words: "And yet it moves" is a legend and nothing else. There is no truth in it.

The Austrian philosopher, inventor of the so-called philosophical Relativism, Paul Feyerabend, wrote in 1976 with respect to Galileo, that the man had been no victim of medieval obscurantism. He declared that at that time the church had been much closer to reason than Galileo himself by taking into account any social and ethical consequences of the quarrel. Feyerabend thought that the verdict had been rational and just. The physicist Carl Friedrich von Weizsaecker went even further. His conclusion is that Galileo was travelling along the path that led directly to the atomic bomb, since Galileo tried to stand for a science without any bounds at all. The Inquisition did not want to let that pass.

In the end, the Vatican conceded that there had been "mistakes" on their part at that time, that both the Pope and some members of the Curia did not support the outcome of the proceedings. Suddenly, for the church Galileo became a man of high esteem, possessed of an exploratory urge to look at nature as a book of God. But it is still very difficult to disarm the myth surrounding this whole affair that has accumulated over centuries. It did not even help in 1741 when Pope Benedict XIV allowed the printing of Galileo's complete works without any abridgement.

The church's opinion of the old genius today is somewhat subdued. Cardinal Brandmueller, long time president of the Papal Committee for the Historical Sciences, still considers Galileo to be a vain self-absorbed scholar who sometimes overdrew his account and certainly showed no inhibitions in the treatment of his colleagues and competitors. Brandmueller says that the verdict had been well founded, firstly, because Galileo had obtained the print permission of the Dialogo by fraud, and secondly—again—because he had been unwilling to present his theory as a pure hypothesis and not as the exact description of reality. He believes that the real scientific importance of Galileo lies in his last major work, presented in the Discorsi, on which he started to work while in his "dungeon" in the palace of the Divine Office, while profiting from the best cuisine in Rome.

Chapter 11
Time Line

Introduction

The Table 11.1 gives a time line encompassing events from the history of science or natural history relevant to our subject and treated here. Included are detailed incidents from the life of Galileo. The scale of this time line is of no relevance.

Table 11.1 Time line of important events in the history of science pertinet to the achievements of Galileo

490 BC	Birth of Anaximander
384 BC	Birth of Aristotle
310 BC	Birth of Aristarchus
287 BC	Birth of Archimedes
273 BC	Birth of Eratosthenes
100	Birth of Ptolemy
150	The Almagest
165	The Canon of the Holy Scriptures
980	Birth of Avicenna
1343	Foundation of the University of Pisa
1401	Birth of Nikolaus Cusanus
1466	Birth of Erasmus of Rotterdam
1473	Birth of Copernicus
1475	Birth of Michelangelo Buonarroti
1483	Birth of Martin Luther
1492	Discovery of America
1509	Birth of John Calvin

(continued)

© Springer International Publishing AG, part of Springer Nature 2018
W. W. Osterhage, *Galileo Galilei*, Springer Biographies,
https://doi.org/10.1007/978-3-319-91779-5_11

Table 11.1 (continued)

1529	Siege of Vienna
1530	Birth of Ivan the Terrible
1533	Birth of Elizabeth I
1537	The Medici return to power in Florence
1542	Birth of Mary Stuart
	Initiation of the Inquisition by Pope Paul III
1543	Publication of "De revolutionibus orbium coelestium"
1545	Beginning of the Counter-Reformation
1546	Birth of Tycho Brahe
1547	Birth of Miguel Cervantes
1548	Birth of Giordano Bruno
1559	First edition of the Index Librorum Prohibitorum
1562	Beginning of the French Wars on Religion
1564	*Birth of Galileo in Pisa*
	Birth of William Shakespeare
1571	Birth of Johannes Kepler
1572	*Relocation of Galileo to Florence*
	St. Bartholomew's Day
1575	Borro: "De motu gravium et levium"
1578	*Galileo in Maria de Vallombrosa*
1580	*Galileo starts his studies in medicine in Pisa*
1583	First Colony in Canada
1585	*Galileo starts giving tutorials*
1587	*Galileo: "Theoremata circa centrum gravitates solidorum"*
1588	*Galileo: "The Topography of Dante's Hell"*
	Galileo: "La Bilancetta"
	Defeat of the Spanish Armada
1589	*Galileo obtains chair in mathematics at Pisa University*
1592	*Galileo finishes "De motu antiquiora"*
	End of Galileo's contract
1596	Birth of René Descartes
1598	End of the French Wars on Religion
1599	*Prolongation of Galileo's assignment in Padua*
1600	Execution of Giordano Bruno
1603	*Extension of Galileo's contract in Padua*
1604	Appearance of a Super Nova
1605	*"Dialog about the New Star" by Ceccio di Ronchitti, a pseudonym*

(continued)

Table 11.1 (continued)

1608	Lippershey invents a telescope
1609	*Galileo constructs his Occhiolino*
	Galileo demonstrates his telescope
1610	Thomas Harriot detects sunspots
	Galileo detects Jupiter's satellites, observes sunspots
	Galileo: "Siderius Nuncius"
	Appointment of Galileo as First Mathematician by the Grand Duke of Tuscany
1611	*Galileo in Rome, appointment to the Accademia dei Lincei*
	Kepler describes his telescope in "Dioptice"
1616	*Instruction from Pope Paul V against Galileo*
1618	Beginning of the Thirty Years War
	Discovery of three different comets
1619	Publication of "Joannis Kepleri Harmonices Mundi Libri Quinque"
1621	*Galileo elected Consul of the Accademia Fiorentina*
1623	*Galileo: "Il Saggiatore"*
1624	*Galileo meets Pope Urban VIII in Rome*
1625	New Amsterdam
1629	Birth of Christiaan Huygens
1630	The plague in Italy
1632	*Galileo: "Dialogo di Galileo Galilei sopra i due Massimi Sistemi del Mondo Tolemaico e Copernicano"*
1633	*Galileo's Renunciation before the Inquisition*
	Galileo in Arcetri
1635	*Galileo: "Discorsi e Demonstrazioni Mathematiche intorno a due nuove scienze", published by Elsevier in Latin*
1641	*Replacement of Viviani by Torricelli*
1642	Birth of Isaac Newton
	Death of Galileo
1648	End of the Thirty Years War
1654	Viviani publishes his biography of Galileo
1672	Beginning of the Dutch War
1687	Newton: "Philosophiae Naturalis Principia Mathematica"
1798	Cavendish' gravitational experiments
1842	Robert Mayer formulates the First Law of Thermodynamics
1854	*Publication of "De motu antiquiora" by Eugenio Alberi*
1865	Rudolf Clausius formulates the Second Law of Thermodynamics
1879	Birth of Albert Einstein

(continued)

Table 11.1 (continued)

1881	Michelson experiment
1889	Birth of Edwin Hubble
1905	Theory of Special Relativity
1906	Eoetvoes' gravitational experiments
1915	General Theory of Relativity
1942	Birth of Steven Hawking
1964	Gravitational experiments by Roll, Krotkov and Dicke
1967	Touch down of Venera 4 on Venus
1985	Foundation of ZARM
1990	Placement of the Hubble Telescope
1992	*Rehabilitation of Galileo*
2008	*Opening of Galileo's trial records*
2009	Placement of the Kepler Telescope
2012	Discovery of the Higgs Boson

References

1. P. C. Haegele: "Das kosmologische anthropische Prinzip", Universität Ulm, Fachbereich Physik, Kolloquium für Physiklehrer am 11. Nov. 2003
2. J. P. Wolfers (Ed.): "Sir Isaac Newton's Mathematische Principien der Naturlehre", Berlin, Verlag von Robert Oppenheim, 1872
3. M. H. Shamos (Ed.): "Great Experiments in Physics", Dover Publications Inc. New York, 1987
4. https://www.leifiphysik.de/mechanic/freier-fall-senkrechter-wurf/geschichte/die-untersuch-ung-des-freien-falls-durch-galilei
5. M. Camerota and M. Helbing: "Galileo and Pisan Aristotelianism: Galileo's 'De Motu Antiquiora' and the Quaestiones de Motu Elementorum of the Pisan Professors", in Early Science and Medicine, Vol. 5, No. 4, 2000, pp. 319-365, Brill
6. C. W. Misner et al.: "Gravitation", W. H. Freeman & Co. New York, 1973
7. J. Mansfeld: "Die Vorsokratiker", Phillip Reclam jun., Stuttgart, 1999
8. K. Manitius: "Des Claudius Ptolemäus Handbuch der Astronomie", Teubner, Leipzig, 1912
9. http://www.webexhibits.org/calendars/year-text-Copernicus.html
10. L. Leiner and J. Boxheimer: Barockprojekt 2000, Mannheim, Juni 1997
11. D. Herrmann: "Die antike Mathematik", Springer Spektrum, Heidelberg, 2014

Index of Persons

A

Aetios, 71
Alberi, Eugenio, 36
Ammannati, Giulia degli, 20
Anaxagoras, 70
Anaximander, 39, 41, 42, 70, 71, 73, 104
Anaximedes, 70
Apelles, 54, 55
Apollodor, 70
Apollonius, 75, 76
Archimedes, 17, 22, 24, 31, 37, 126, 136
Aristarchus, 73
Aristotle, 17, 19, 21, 33, 36, 38, 59, 65, 68, 69,
    74, 108, 109, 122, 125, 129, 134
Arnauld, Antoine, 134
Arsini, Ranieri, 30
Averroes, 37
Avicenna, 37, 40

B

Bacon, Roger, 74
Baliani, Giovan Battista, 56, 66
Ballestra, Francesco, 118
Bandini, Ottavio, 53
Barberini, Fransesco, 115, 120, 123, 129
Bardi, Girolamo, 122
Barton, Catherine, 135
Beeckmann, Isaac, 134
Bellarmino, Roberto, 54, 58, 59, 118, 119, 140
Benedetti, Giovan Battista, 137
Benedict XIV, 141
Bernegger, Matthias, 123, 125, 127
Bohr, Niels, 65
Bolognetti, Giorgio, 129
Borghini, Jacopo, 20

Borro, Girolamo, 31, 36, 37
Brahe, Tycho, 57, 61, 64, 76, 84, 88
Brandmueller, Walter, 141
Brecht, Berthold, 138
Brunelleschi, Filippo, 80
Bruno, Giordano, 14, 40, 42, 136
Buonamici, Francesco, 31, 36–38
Buonarotti, Michelangelo, 12, 33, 80, 129
Buridanus, Johannes, 137

C

Caccini, Tommaso, 58
Calvin, John, 11, 123, 132
Campanella, Tommaso, 56, 115
Castelli, Benedetto, 57, 114, 123
Catherine of Aragon, 10
Cavendish, Henry, 30
Cervantes Saavedra, Miguel de, 13, 86
Cesare, Alberto, 127
Cesi, Federico, 58, 66, 137
Charles I, 92
Ciampoli, Giovanni Battista, 114
Cioli, Andrea, 119, 120
Clausius, Rudolf, 126
Clavius, Christoph, 53
Clemens VI, 30
Colombe, Lodovico delle, 58
Columbus, Christophe, 39, 74, 79
Compte, Auguste, 40
Copernicus, Nicolas, 3, 42, 46, 56, 57, 59, 75,
    77, 79, 80, 83, 113, 114, 118, 132, 140
Cortes, Hernan, 79
Cosimo II, 48
Cromwell, Oliver, 92
Cusanus, Nikolaus, 79, 85, 132

© Springer International Publishing AG, part of Springer Nature 2018
W. W. Osterhage, *Galileo Galilei*, Springer Biographies,
https://doi.org/10.1007/978-3-319-91779-5

Subject Index

Printed in the United States
By Bookmasters